CW00613216

Amateur Radio
INSIGHTS

Work the World
with DMR

Digital Mobile Radio Explained

by

Andrew Barron, ZL3DW

This book has been published by Radio Society of Great Britain of 3 Abbey Court, Priory Business Park, Bedford MK44 3WH, United Kingdom
www.rsgb.org.uk

First Edition 2022

Amateur Radio Insights is an imprint of the Radio Society of Great Britain

ISBN: 9781 9139 9518 8

Cover design: Kevin Williams, M6CYB
Typography and design: Andrew Barron ZL3DW
Production: Mark Allgar, M1MPA

Printed in Great Britain by Short Run Press Ltd. of Exeter, Devon

Any amendments or updates to this book can be found at:
www.rsgb.org/booksextra

A note from the author

This book is a guide to Digital Mobile Radio (DMR). It will help you to become familiar with the complex terminology used by the DMR crowd, purchase and program a DMR radio, and make your first few calls. It also includes sections on how to configure and use a DMR hotspot, using DMR repeaters, and talk groups. The book takes a practical approach, providing the information you need to get started with this exciting digital voice technology. Before you know it, you will be able to talk with amateur radio operators all over the world.

Many of the examples in the book are based on my observations using the DMR equipment that I own. There will be variations between the equipment that I am using and your equipment, and that may mean that some things mentioned in the book don't work or need a minor adjustment. But most of the information is generic and will apply to all DMR radios, hotspots, and services. Where possible I will try to cover any variations. The rules regarding your local repeaters and DMR groups may be different, so it is advisable to talk with members of your local radio club and the DMR community.

Work the world with DMR

Other books by Andrew Barron

Testing 123
Measuring amateur radio performance on a budget

Amsats and Hamsats
Amateur radio and other small satellites

Software Defined Radio
for Amateur Radio operators and Shortwave Listeners

An introduction to HF Software Defined Radio
(out of print)

The Radio Today guide to the Yaesu FTDX10

The Radio Today guide to the Yaesu FTDX101

The Radio Today guide to the Icom IC-705

The Radio Today guide to the Icom IC-7300

The Radio Today guide to the Icom IC-7610

The Radio Today guide to the Icom IC-9700

ACKNOWLEDGEMENTS

Thanks to the DMR network operators who develop and maintain the DMR talk groups and IT infrastructure. They have created first-class worldwide networks and made them available to all licensed amateur radio operators to use for free. Also, a huge thankyou to all the hardware and software developers who worked for thousands of hours, often for free, to adapt DMR technology for amateur radio use. And finally, many thanks to you, for taking a chance and buying my book.

ACRONYMS

The amateur radio world is chock full of commonly used acronyms and TLAs (three-letter abbreviations :-). They can be very confusing and frustrating for newcomers. I have tried to expand out any unfamiliar acronyms and abbreviations the first time that they are used. I have assumed that anyone buying a DMR radio will be familiar with commonly used radio terms such as repeater, channel, MHz, and kHz. Near the end of the book, I have included a comprehensive glossary, which explains many of the terms used throughout the book. My apologies if I have missed any.

Approach

There is a huge amount of excellent information about DMR (Digital Mobile Radio), in the form of online websites, forums, pdf files, and "how to do it" videos. It is not my intention to replace these valuable resources. I hope to complement the insights that they provide and concentrate as much information as possible into a single document. The radios used as examples in this book are the Radioddity GD-AT10G which is very similar to several AnyTone, Alinco, and BTECH models, and the TYT MD-UV380 which is a popular model. That also covers the TYT MD-UV390 and to some extent Retevis radios and other TYT models including the MD380 and MD390.

The book is a practical guide that explains the steps that you need to follow to make your new DMR radio work on your local repeater or hotspot, and for worldwide contacts. It is not as simple as entering a couple of frequencies and setting a CTCSS tone the way you would for an FM radio. There are a lot of new terms to discover, including dashboards, zones, receive groups, colour codes, code plugs, hotspots, Parrot, talk groups, and time slots. Also, acronyms like MMDVM, CPS, IPSC2, DMR-MARC, TGIF, and DMR+.

Amateur radio DMR is not easy. At least not if you want to configure the radio yourself and program a hotspot. There are shortcuts if you only want to use a single repeater with a few talk groups, or if you can get a ready-made 'code plug' (radio configuration file) for your radio. But to get the most out of the amazing features that DMR offers, such as worldwide amateur radio contacts over several DMR networks, you can expect a steep learning curve. That's where this book will be the most helpful.

MMDVM (multi-mode digital voice modem) 'hotspots' are very popular accessories. I have included information about their uses and configuration. I cover the new duplex hotspots and the more familiar simplex hotspots, including a section on how to assemble a hotspot from a kit, a Raspberry Pi, and an SD card. This is followed by step by step instructions for configuring the Pi-Star hotspot operating system.

I also included some information on the DMR radio data structure and some observations on the advantages and disadvantages of the DMR technology over FM, and other digital voice modes such as System Fusion, D-Star, and P25.

Many readers will already have bought a DMR handheld or mobile, but for those that haven't, there is a brief guide to the features that you should look for. I went for a mid-range option, hoping to get a good set of features at a reasonable price. A few very cheap DMR handhelds are not suitable for amateur radio because they don't support the DMR Tier II standard with two DMR time slots.

The Glossary explains the meaning of the many acronyms and abbreviations used throughout the book and the Index is a great way of getting to a specific topic.

Conventions

The following conventions are used throughout the book.

Mobile Station (MS). A 'mobile' usually refers to a vehicle-mounted radio. DMR is based on commercial radio systems, so most technical standards refer to 'mobile' radios and 'base station' radios. In this book, 'mobile' means a vehicle-mounted radio or a handheld radio.

Base Station (BS). A base station is the commercial radio term for a DMR repeater.

Fusion or System Fusion is the digital voice standard employed by Yaesu radios. It is often abbreviated to YSF (Yaesu System Fusion).

D-Star is the digital voice standard employed mostly by Icom radios. Unlike DMR, it was developed specifically for amateur radio use.

DMR is an ETSI (European Telecommunications Standards Institute) digital voice standard employed by a wide variety of manufacturers. Motorola and Hytera are the biggest commercial vendors of DMR radios and trunked radio repeater systems. Motorola calls their version of DMR, MotoTRBO.

CPS (Customer Programming Software) is a daft name for the PC radio configuration software. But it is in widespread use, so I followed the trend and used CPS as an abbreviation for the radio configuration software on your computer.

I have used a **different font** to indicate menu steps in the CPS software and data entry using the radio buttons.

I use the > symbol to show steps in a menu structure. For example,
Common Setting > Optional Setting > Work Mode.

I use the ↳ and → symbols to show choices in a menu structure. For example,
Common Setting
      ↳ Channel → Channel Name
      ↳ Zone
      ↳ Scan List

'**Click**' or '**left click**' means to click the left mouse button. '**Right click**' means to click the right mouse button. 'Click on,' means to hover the mouse over a button or menu option on the PC software, and then click the left mouse button to make the selection.

What is DMR?

DMR stands for Digital Mobile Radio. It was developed as a way of reducing the bandwidth of the transmitted signal while improving the quality of the received voice transmission. DMR can transmit two voice channels on the same RF carrier, using the same 12.5 kHz bandwidth as a single FM repeater channel. Also, the digital voice signal is less prone to noise and flutter fading, resulting in excellent audio quality. This does come at the expense of needing a slightly higher received signal strength than an FM signal. Typically, the DMR signal from a repeater will sound "perfect" or it won't be received at all. On Internet-linked services such as amateur radio DMR talk groups, packet loss or delay in the Internet traffic can sometimes break up or distort the audio, but this is not a failure of the DMR link.

In 2005 the Digital Mobile Radio Association was formed to promote compliance to the ETSI (European Telecommunications Standards Institute) standards among its member manufacturers. The association now includes more than 160 DMR equipment manufacturers. Motorola was the first company to market a range of DMR radios. These days, Hytera and Motorola are the largest producers of DMR equipment. Motorola uses the term MotoTRBO in place of DMR.

The biggest advantage of DMR systems is the 'open ETSI standard' which ensures that radios from all DMR radio manufacturers can be used on any DMR network. This is especially important for amateur radio DMR because it means you can buy any DMR Tier II radio, and it will work on your local DMR repeater or any DMR hotspot.

DMR supports private calling, group calls, short messages (SMS), GPS location, error correction, talk groups, better battery life than FM radios, and man-down safety features. Some DMR radios and networks also support APRS. The dual time slots make DMR difficult for scanners to decode, with the option to add encryption on Tier II and Tier III radios and network encryption on Tier III trunk systems. However, in most countries, encryption is not legal for amateur radio systems.

References: https://cwh050.mywikis.wiki/wiki/MotoTRBO
https://www.hytera.us/resources/dmr-tier-iii-3

The biggest disadvantage of DMR is that the system was developed for commercial radio networks where a user is given a radio that is pre-programed with the channels and talk groups for the network. Although you can select or add talk groups, channels, and zones using the buttons on the radio, it is rather tedious. Realistically, to change the configuration, a new 'code plug' configuration file must be downloaded into the radio. You need a channel for every talk group that you want to use, on every repeater or hotspot that you want to use it on. And each channel must be in a zone, or you cannot select it. You must also get the time slot and the colour code right. Programming your radio is very complicated. It takes a lot of time to get it right.

TECHNICAL INFORMATION

The ETSI standard for DMR equipment includes three 'Tiers' which denote increasing levels of capability while maintaining downwards compatibility. A Tier III radio can operate in the Tier II mode and a Tier II radio can operate in the Tier I mode. All DMR radios are also capable of operating in ordinary FM mode.

DMR Tier I was originally intended for 'unlicenced' use on the 446 MHz 'public radio service' band, (CB radio). But now, Tier I radios are also available for the VHF and UHF amateur radio bands. Some of the cheapest DMR handheld radios being sold online are Tier I. These radios should be avoided because they are not able to use two time slots and cannot be used on amateur radio DMR repeaters. They are only good for direct radio to radio contacts.

DMR Tier II radios are often marketed as having two time slots or 'dual time slots' and the listing or advertisement may also state compliance with DMR Tier II or MotoTRBO Tier II. Most are dual-band, covering the 2m and 70cm amateur bands. This is not of critical importance because almost all DMR repeaters are on the 70cm band. DMR Tier II is the standard for amateur radio DMR.

There is no point in paying extra for a Tier III radio. DMR Tier III radios are designed for trunked radio systems which include one or more multi-channel repeaters and usually a manned 'operations' or 'dispatch' centre. They run a continuously 'on' control channel using one time slot of one of the RF channels. This regularly polls the DMR radios within the repeater coverage area and supplies status and often location information back to the 'Central Office.' When you make a 'call' the data goes from the repeater over a private data link to the MSO ('Mobile Switch Office'). From there the call is routed to the operator, another radio, or sometimes a telephone extension. There are facilities for recording all calls on the network and for end-to-end encryption. If you called a radio that is currently using the same repeater, the system can optionally change the call to a direct connection through the repeater which does not go back through the MSO. Amateur radio is not suited to this 'trunk radio' type of operation so it is unlikely that Tier III will ever be introduced to our DMR networks.

DMR COVERAGE COMPARED TO FM ANALOG COVERAGE

When operating outside, you can expect 'line of sight' communications, like all radio transmissions on the 70cm band. If you can see the hill or building the repeater is on, you should be able to use the repeater. The situation is different when you are inside a building, in a built-up area, or woodland etc. Again, you can expect the DMR coverage to be similar to that of an FM repeater. On the fringes, DMR generally has superior coverage performance to analogue FM because the forward error correction used in the DVSI AMBE+2 digital voice CODEC can cope with bit error rates as high as 5% with no degradation in the perceived speech quality.

DMR RADIOS TO AVOID

The following radios are fine for direct radio to radio, simplex contacts, but most are DMR Tier I radios that cannot be used on DMR repeaters. The others have poor reviews. Avoid the Baofeng DM5R, DM5R+, RD-5R. Buy the DM-1801 or one of the other DMR Tier II modes. Also avoid the Radioddity GD55, GD55+, and GD77. The Retevis RT82, Zastone Mini9+, TYT MD-398, and the MD-680. Buy a Tier II model such as the Radioddity GD-AT10G or TYT MD-UV380. I'm sure that there are many other duds out there, particularly on the second-hand market.

BUYING A DMR

Here are some things to consider.

- Handheld or mobile. What suits your operating style and budget?

- Dual-band (UHF/VHF). Most DMR repeaters are on the UHF 70 cm band, but a dual-band radio may be better if you are using the FM mode.

- **Must** be 'Tier II' or 'dual time slot'

- Large User ID 'Contacts' database. If you want most calls to display a callsign and name rather than the DMR ID number or 'Unknown ID'

- Programming cable and charger supplied. An extra cost if it is not

- Colour screen that is easy to read and provides plenty of information

- Waterproof, IP67 rating or similar if you are planning on hiking or mountain climbing

- Good CPS radio configuration software. This can save you hours of frustration. The CPS for the TYT MD-380 and MD-390 is terrible. The CPS for the TYT MD-UV380 and MD-UV is better, and the CPS for the Radioddity GD-AT10G, AnyTone AT-D878UV, Alinco DJ-MD5T, and BTECH DMR-6X2 is very good.

- Good online reviews. These are always subjective, but they can identify a real stinker. I tend to read the negative one-star and two-star comments.

VOICE CODING

When you transmit, your voice is converted into a digital data stream of binary bits which is applied to the 4FSK modulator in the transmitter. In the receiver, the recovered data stream is converted back into an audio signal for you to hear. The conversion is done with a propriety chip running the DVSI (Digital Voice Systems Incorporated) AMBE+2 Vocoder. A Vocoder is a CODEC (coder-decoder) designed for voice signal digital coding, compression, multiplexing, and encryption.

The older AMBE Vocoders are used for Inmarsat and Iridium satellite phones, D-Star, and 'phase 1' versions of P25 digital. DMR, including Motorola's MotoTRBO, System Fusion, NXDN, and 'phase 2' P25, use the later AMBE+2 Vocoder.

Software digital voice programs like FreeDV usually use the open-source Codec2 software, which is free and uses half the bandwidth of AMBE to encode speech at a similar voice quality.

MODULATION

DMR Tier II uses 'four state frequency shift keying' (4FSK). Each frequency shift or 'symbol' carries two bits of the digital data stream. The RF signal fits into a 12.5 kHz channel, but it carries two voice channels, making DMR more spectrally efficient than D-Star, Fusion, or P25. The effective data transmission rate is 4.8 kbs per voice channel. The emission designator is 7K60FXE, indicating that DMR is an FM mode for 'voice' transmissions that uses 7.60 kHz of bandwidth.

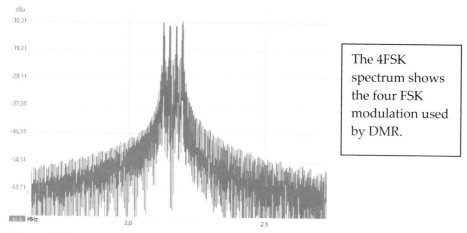

The 4FSK spectrum shows the four FSK modulation used by DMR.

Figure 1: The RF spectrum of a 4FSK signal.

https://www.researchgate.net/figure/Spectrum-of-4-FSK-signal_fig1_261158152

The FSK frequencies are on the channel frequency +1944 Hz, +648 Hz, -648 Hz, and -1944 Hz. The C4FM modulation used for System Fusion and some P25 versions also use four frequencies but the offset is ± 600 Hz and ± 1800 Hz.

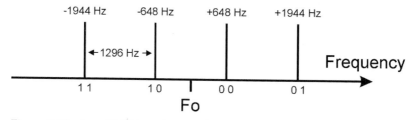

Figure 2: The four DMR tone frequencies

THE DIFFERENCE BETWEEN FSK AND PSK

With frequency shift keying (FSK) the binary data signal is represented by changes in the RF frequency. In phase-shift keying (PSK) the frequency remains the same and the binary data is represented by changes in the phase of the signal. Each phase or frequency state is known as a symbol. The symbol rate for 2FSK or PSK is the same as the bit rate because there are only two possible frequency or phase states, each representing one binary bit. Note that it is not the current frequency or phase that determines whether the symbol is a binary one or a zero. It is the change of state. There are scrambling and data coding techniques aimed at eliminating long runs of ones or zeros. Many systems use differential coding where a change of state means that the current binary bit is different to the previous bit. No change of state indicates that the current binary bit is the same as the previous bit.

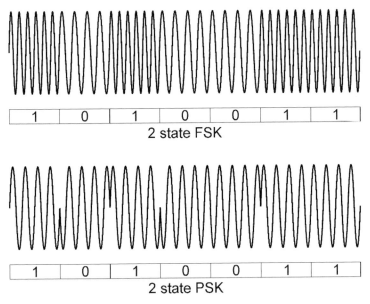

Figure 3: Two frequency FSK vs Binary PSK modulation

4FSK (four frequency shift keying) uses four frequencies rather than two frequencies. 4PSK (four-phase shift keying) also known as QPSK (quadrature phase-shift keying) uses four phase changes, 0°, 90°, 180°, and 270°, rather than the 0° and 180° phase shifts used in 2PSK which is also known as BPSK (binary phase-shift keying).

With 4FSK and QPSK, each phase or frequency state carries 2 bits of the digital signal, so the bit rate is twice the symbol rate. Using this higher order of modulation allows you to transmit twice as much data in the same amount of time. The data is usually manipulated so that all four states are being transmitted regularly. This is particularly important for QPSK because it affects the shape of the transmitted spectrum and clock recovery at the receiving station.

THE DATA FORMAT

DMR is a two-channel TDMA system. TDMA stands for time division multiple access. It means that the radio 'base station' or repeater sends some data for one voice channel and then some data for the other voice channel and keeps on repeating that pattern. Because the system sends the voice channels at consecutive times, each voice channel is called a **time slot**. Usually abbreviated to TS1 and TS2.

A repeater always transmits two time slots. A mobile or handheld radio transmits on one voice channel or the other, so the transmitter sends data in the allocated time slot, then turns the transmitter off during the unused time slot. Another operator transmitting on the other time slot can use the repeater at the same time. The mobile transmitter only transmits 50% of the time and this extends the battery life. The time slots are offset. The mobile either transmits on TS1 while the repeater is sending data for TS2. Or it transmits on TS2 while the repeater is sending data for TS1.

The DMR transmission data structure is independent of the frame structure of the data being transmitted. For example, the Vocoder may use a 20 ms data frame, while the DMR uses a 30 ms data burst. Or the radio may be carrying some other data such as text or digital images. It does not matter what data structure the DMR transmission is carrying, provided the overall transmission is fast enough to ensure that the 'payload' data arrives at the destination at a rate that allows the original data frames to be recreated.

While it is transmitting 'voice traffic' the DMR data structure is comprised of 30 ms data packets called 'bursts' which contain the data for one voice channel, plus sync bits, and embedded signalling data. Two bursts make a TDMA Frame. Each 60 ms TDMA Frame carries data and voice bits for both time slots. The repeater 'base station' also transmits some CACH (Common Announcement Channel) bits after each TDMA Frame which contain the TDMA Frame number and low data rate signalling. The CACH information also carries the bits that tell a mobile whether the repeater time slot is in use. The mobile will not transmit unless the CACH data says that the time slot is idle. The CACH data sent after TS1 tells the mobile whether TS2 is available and the CACH data sent after TS2 tells the mobile whether TS1 is available.

The overall data rate is 9600 bits per second.

- 3600 bps for the TS1 voice channel
- 800 bps for TS1 sync and signalling data
- 3600 bps for the TS2 voice channel
- 800 bps for TS2 sync and signalling data, and
- 800 bps for CACH bits

The data sent from a base station (repeater) is different to the data sent from a mobile (mobile or handheld) radio. The format varies depending on whether the radio is transmitting data or voice traffic. The base station transmits continuously while there is data to send.

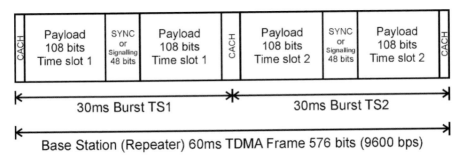

Mobile data for 'voice traffic' is sent during one time slot or the other, so there are gaps between each burst. Ah-ha! That's why they are called bursts!

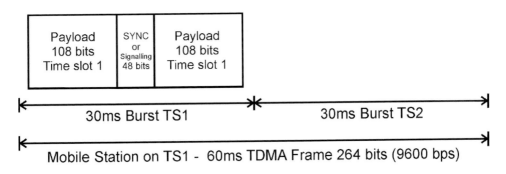

A mobile station transmits on TS1 or TS2. It does not send CACH bits, so the actual transmission time is 27.5 ms. The 5 ms between TS1 and TS2 allows time for the transmitter to ramp up to full power and for any minor timing issues between radios.

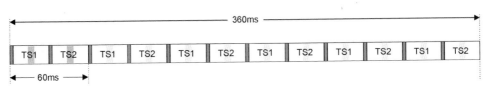

Figure 4: A Base Station 'voice' superframe

Repeater voice traffic is sent in 360 ms 'superframes' of 12 bursts (6 TDMA frames). The first burst on each time slot carries 216 bits of digital voice data and 48 bits of 'voice sync' data. The remaining ten bursts carry 216 bits of digital voice data and 48 bits of signalling data. Then the pattern repeats.

Figure 6: A 'voice' superframe from a mobile station transmitting on TS2

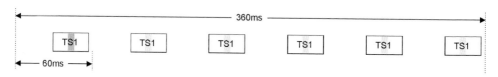

Figure 5: A 'voice' superframe from a mobile station transmitting on TS1

A mobile or handheld radio transmits one burst with 216 bits of voice data and 48 bits of 'voice sync' followed by five bursts with 216 bits of voice data and 48 bits of digital signalling data. After 360 ms the pattern repeats. If required, two bursts of data traffic such as GPS data or SMS text can be inserted before each 360 ms voice superframe. The 5 ms gap before the voice data begins allows time for the transmitter to ramp up to full power and for any minor timing issues between radios.

The data rate

A mobile, handheld, or simplex hotspot, transmits 216 bits of voice data and 48 sync or data bits in one 30 ms burst every 60 ms. The data is sent at the 9600 bps rate during the burst resulting in an average data rate of 4400 bits per second.

A repeater transmits 576 bits every 60 ms. That is a data rate of 9600 bits per second or 4800 bits per second per time slot.

DMR Repeater	
Bits per 60 ms TDMA Frame	
12 bits	CACH
108 bits	TS1 voice
48 bits	Sync or data
108 bits	TS1 voice
24 bits	CACH
108 bits	TS2 voice
48 bits	Sync or data
108 bits	TS2 voice
12 bits	CACH
576 bits	TOTAL

DMR Repeater	
Effective data rates	
TS1 voice (216 bits)	3600 bps
Sync or data (48 bits)	800 bps
TS2 voice (216 bits)	3600 bps
Sync or data (48 bits)	800 bps
CACH (48 bits)	800 bps
TOTAL	9600 bps

MOBILE OR HANDHELD SIMPLEX VS REPEATER MODE

If your Radioddity/AnyTone radio is using a channel set to DMR mode = **Repeater** and there is a split between the receive and transmit frequencies, the radio will send a data burst every 550 ms to the repeater (or duplex hotspot). After it sees an acknowledgement back from the repeater it starts to send 30 ms TDMA voice data bursts as discussed in the previous section.

If the channel is set to DMR mode = **Simplex** or the transmit and receive frequencies are the same, the radio sends 30 ms TDMA voice data bursts as discussed in the previous section. It does not wait for an acknowledgement from the hotspot, which is why a simplex hotspot is faster when you go from receiving to transmitting than a duplex hotspot or a repeater. If the frequencies are the same, the radio will operate in the simplex mode irrespective of the DMR mode setting. The CPS says to use Repeater for a simplex hotspot, but I don't think it matters.

If the channel is set to the **Double Slot** DMR mode, the radio will send Simplex data bursts on the nominated time slot. It means that you can use the same frequency for two direct radio-to-radio conversations. One pair of radios can use TS1, and at the same time, another pair of radios can use TS2.

The TYT CPS does not have a setting for 'DMR mode,' but it behaves the same way. If the frequencies are both the same, it works in simplex mode and if they are different it works in repeater mode.

TIP: There is some confusion on online forums about this, but I have checked with a fast oscilloscope, and this is what happens.

COMPARISON OF DIGITAL VOICE MODES

Some people say that DMR has superior voice quality because it uses the more advanced AMBE+2 Vocoder. I believe that any differences are subjective, and all three digital voice systems provide excellent speech quality. The main issue with audio quality is not the digital radio component but packet loss in the Internet routing that links the repeaters and hotspots to the world. I guess there are a thousand reasons why your experience might vary.

All the digital voice systems in common use on the UHF and VHF amateur radio bands use digital forms of FM modulation. This is because it has been traditional to use FM at those frequencies and because the commercial operators are interested in migrating customers from FM repeaters onto digital repeaters. It is good because it means that all DMR radios also support FM operation and can access FM repeaters.

The 4FSK modulation used for DMR and C4M modulation used for YSF produces a four state constellation the same as QPSK but without the phase component, and the bit error rate (BER) performance is nearly identical (within about 1 dB).

The GMSK modulation used for D-Star is slightly better. DMR requires a received signal two or three dB stronger to achieve the same BER as D-Star.

All of the modes in this comparison use extra transmitted bits to perform FEC (forward error correction) which improves the voice quality by repairing errors in the received data.

RF factors	DMR	D-Star	Fusion	P25 phase1
Vocoder	AMBE+2	AMBE	AMBE+2	IMBE
FEC	Yes	Yes	Yes	Yes
Modulation	4FSK	GMSK*1	C4FM*2	C4FM*2 or CQPSK
Emission	7K60FXE	6K00F7W	9K36F7W	8K10F1E
Transmission rate per voice channel	4800 bps	4800 bps	9600 bps	9600 bps
Mod Bandwidth	7.6 kHz	6.0 kHz	9.36 kHz	8.1 kHz
Channel bandwidth	12.5 kHz	6 kHz	12.5 kHz	12.5 kHz
Voice channels	2	1	1	1
Developer	ETSI	JARL	Yaesu	APCO

User information	DMR	D-Star	Fusion
Registration required	Yes	Yes	No
User ID transmitted	DMR ID*3	Callsign	Callsign
User ID displayed	DMR ID*4	Callsign	Callsign
Is the ID OK for FCC and other jurisdictions?	No	Yes	Yes

*1 C4FM (continuous envelope 4 frequency modulation) is a variant of 4FSK used by Yaesu System Fusion and P25 phase 1.

*2 GMSK (Gaussian Minimum Shift Keying) is a variant on QPSK used for D-Star.

*3 most amateur radio licencing authorities require you to transmit your callsign. DMR does not do this automatically. You need to identify your transmission by saying your callsign, the same as you would on an FM repeater.

*4 the mobile, handheld, or hotspot will only display the other station's callsign if it is held in the contact list in your radio, or the Talker Alias includes the callsign. Otherwise only the ID number is displayed. There are around 206,000 registered DMR users. Some radios can store all of them. With other radios, you will have to be more selective. You can download selected data from https://www.radioid.net/. Or you may elect to only enter contact data from your local area, or for Hams you call often.

User repeater connections	DMR	D-Star	Fusion
Talk to local stations	Yes	Yes	Yes
Link to another repeater	No	Yes	No
Internet-linked systems	Talk groups	Reflectors	Wires-X rooms
Selection	Channel switch	UR field	Room name
Group Call	Yes	Yes	Yes
Private Call	Yes	Yes	No
Echo test	Yes	Yes	No

Ease of use	DMR	D-Star	Fusion
Memory selection	Channel switch	Dial or list	Channel switch
Analog mode	Programmed channel	Button	Button
Programming	Difficult*1	Medium	Easy
Using the radio	Easy	Medium	Easy
Radio models (mostly)	Many	Icom	Yaesu
Radio cost	Cheapest	Medium	Most expensive
Hot spots	Any MMDVM Pref Duplex	Any MMDVM	Any MMDVM or Wires-X
Send photos	No	Yes	No
SMS (text) messages	Yes	Yes	No
Send GPS/APRS data	Some models	Yes	No

*1 It's not that the programming difficult to accomplish. You just download a new code plug config file using a USB cable. The annoying thing that makes life a little difficult is that you may have to edit and download a new file any time you want to add a new talk group or repeater to the radio. Adding a new repeater can involve adding many additional lines to the file. It is not as easy as flying or driving to another city and entering the repeater channel frequency and CTCSS tone from the radio keypad the way you would for an FM repeater. Most DMR radios can be programmed with the radio keypad, but it is fiddly.

YAESU SYSTEM FUSION CONFUSION

System Fusion (YSF) has two network possibilities. You can connect to the Wires-X system which is operated by Yaesu. To do that you either need a YSF repeater that has a Wires-X interface, or a second radio connected to a Yaesu HRI-200 Wires-X modem which together acts as a hotspot. This is an expensive use of a Yaesu radio, and you must use the Yaesu modem. The alternative is to use a standard YSF repeater or an MMDVM hotspot. This gives you access to some, but not all, of the Wires-X rooms, and some cross-linked talk groups on the DMR or D-Star networks.

Getting started

The first thing you need to do is check out what DMR repeaters are available in your area. It is a good idea to talk to DMR users at your local amateur radio club. If there are no DMR repeaters but you have good a good Internet connection, or you are willing to use your mobile data to provide a WiFi connection, you can buy a hotspot and use that to connect your radio to talk groups all over the world.

Next, you should buy a DMR handheld or mobile radio. Take a look at some things to look out for, on page 5 in the previous chapter. I went for a mid-range Radioddity GD-AT10G. It has four power ranges up to 10 watts, a 3100 mAh battery, a large 200,000 contact list, 4000 channels, a programming cable, GPS and APRS. On the downside, it is UHF only which might not suit you if you want to use the radio on FM repeaters. It also includes radio keypad programming, which can be very handy if you go to a different region. The radio is the same as the AnyTone AT-D878UV except that the AnyTone radio is a dual bander. I also bought a TYT MD-UV380 because it is a popular model, and many other radios are programmed the same way.

Once you have your radio, there is work to do before you can use it. At first, this seems rather daunting because of all the new terms you need to learn and having to download software and databases. But we will step through the process, and it will soon become second nature.

YOUR DMR ID

You cannot use any DMR repeaters or talk groups unless your radio has been programmed with your personal DMR ID number. This number is linked to your amateur radio callsign and will be used for all of your DMR radios and hotspots. There is only one registration source worldwide. It is at https://radioid.net/register or search for 'DMR ID registration.' Note that you **must** be a licenced amateur radio operator to get a DMR ID. Before you start the registration process make sure that you have a copy of your amateur radio licence or operating certificate (1 to 3 pages) in .gif, .jpg, .jpeg, .png, or .pdf format. I had an image from my LoTW application years ago, so that was fine.

TIP: If you buy a used DMR radio or hotspot you must ensure that you change its DMR ID to your DMR ID. Otherwise, you will be effectively illegally using someone else's callsign. If the party you are talking to has a large contact list loaded into their radio, it will display the previous owner's name, location, and callsign.

The process begins with the website asking for your callsign, name, and email address. It then sends you an email which you must open, but you do not need to click a link or reply. I guess it just gets a read receipt from the mail server. Then you can carry on and enter a password for your account. It must be at least 8 characters and contain a minimum of one symbol, one number, and one upper case letter.

Next, a new web page appears. It will usually be pre-loaded with your address. I guess they get that from QRZ.com.

You will have to enter any missing information. I had to put 'South Island' using the State/Prov dropdown list. Save that information to get rid of the red warning bar. The final thing you must do is upload a copy of your amateur radio licence or operating certificate (1 to 3 pages) in .gif, .jpg, .jpeg, .png, or .pdf format.

That makes the second red bar disappear and the application goes into a pending mode while your licence is manually checked. You should receive an email with your DMR ID number within a few days. My application was approved in four and a half hours. I was automatically issued with DMR IDs for my DMR radio and two hotspots.

Your DMR ID is used for all your DMR radios and your first hotspot. Subsequent hotspots get a two-digit 'ESSID' extension 01, 02 etc. to separate them in the network from the first hotspot. A DMR repeater can use your DMR ID with an extension, but most licenced repeaters have a different callsign issued by the licencing authority.

THE MINIMUM POSSIBLE CONFIGURATION

This section covers the most minimal configuration that will get you talking on the air in the shortest possible time. You have just bought a new DMR radio, you are not ready to mess around with CPS software and code plugs, and there is a local DMR repeater that you can use. At this stage, all you need to know about the repeater is its frequencies. Of course, more information will be helpful.

If you **are** ready to learn about the CPS config software and code plug programming, skip this section and jump to 'Download the CPS configuration software' on page 22.

We are going to set up one talk group, on one channel, in one zone. That is the minimum that will work. We are going to do this "the hard way," programming the radio using the keypad buttons. Believe me, if you are programming more than one channel, the CPS or spreadsheet methods are much faster and easier. I will run through the instructions for the Radioddity GD-AT10G which is very similar to several AnyTone, Alinco, and BTECH models, and the TYT MD-UV380 which is a popular model. That also covers the TYT MD-UV390 and probably other TYT models. It is impossible to cover all the models from all of the suppliers. But this should at least provide an idea of the steps that are involved, and some tricks you can use to find out about the talk groups and colour code that the repeater is using.

The Menu (-) button, is the green button on the left just below the display.

The Back (- -) button, is the red button on the right just below the display.

The full menu tree for the Radioddity/AnyTone radios is on page 123.

The full menu tree for the TYT radios is on page 119.

Radioddity/AnyTone etc.

TIP: Don't stop! If you do, your changes will be lost, and you will have to start again.

The radio needs your DMR ID. Nothing will work without that.

Menu > Settings > Chan Set > Radio ID > Add Radio ID > Select > Edit ID, or

Menu > Settings > Chan Set > Radio ID > Select > Option > Edit ID

Edit the DMR ID number. The red 'back' key backspaces. Then **Confirm** with the menu key.

Edit your DMR name the same way. Most people use their callsign – name. For example, ZL3DW – Andrew.

Now we need a talk group. At this stage, we don't know what talk groups are available on the repeater, but one thing is for sure, we will need the 'Local' talk group TG 9. Talk group 9 is used for talking over the repeater without the call extending to the wider network. On many DMR networks, it is also used for talking on 'reflectors,' so if the repeater is linked to a reflector your TG 9 call may be heard over the entire region, state, or country.

Menu > Talk Group > TG List will display a list of the talk groups already on the radio. 'Local' or 'LCL' TG 9 may already be there. The radio will not save two talk groups with the same TG number. To add TG 9, assuming that it is not there already, use Menu > Talk Group > New Contact > Input ID. The display should say Private ID which is not what we want. Click the **#** key to change it to Talk Group ID. Enter a 9 and **Confirm**. Now select **Input Name** and enter **Local** using the keypad buttons. Then click **Confirm**. Finally, scroll down the options and select **Save**. That is all we need for a talk group.

Add a new channel using the radio keypad

- Click the **Menu** (-) button

- Use the round **up-down** button to navigate down to **Settings** and click **Menu** (-) to select.

- Click down to option 2 **Chan Set** and click **Menu** (-) to select.

- On option 1 **New Chan**, click **Menu** (-) to select.

- The radio will select the next available channel number. You can change it if you want to overwrite an existing channel. Then click **Menu** (-) to Confirm

- Enter the channel name **Local TG 9** and click **Menu** (-) to Confirm

- Select a Zone from the displayed list, click **Menu** (-) **Menu** (-) to Confirm and save. There will probably only be one zone available and that's OK for now.

- This creates a channel that is like the channel that the radio was sitting on when you started the process. You will need to carry on to the editing process to change the frequencies, add the talk group, and colour code etc.

Edit the new channel using the radio keypad

- Set the radio to the zone and channel that you want to edit. Since we just added the channel, this should already be correct.

- Click the **Menu** (-) button

- Use the round **up-down** button to navigate down to **Settings** and click **Menu** (-) to select.

- Click down to option 2 **Chan Set** and click **Menu** (-) to select.

- There are 22 settings that you can change. Numbers 3, 7, and 8 are the important ones at this stage.

    3) Channel Type. Set to **D-Digital**

    4) TX power – turbo, **high**, medium, low

    5) Offset – **zero** unless you have a reason to change it

    6) Bandwidth – the only option for digital is 'narrow,' (12.5 kHz)

    7) RX frequency. Enter the repeater output frequency. To exit the screen press P2.

    8) TX frequency. Enter the repeater input frequency. To exit the screen press P2.

    9) Talk around **off**

    10) Name. Was set when you added the channel.

    11) TX Allow **different CC**

    12) TX prohibit **off**

    13) Radio ID. You have already set this

    14) Color Code – select 0 – 16. Set it to **1** unless you know that the repeater uses a different colour code

    15) Time Slot – select **TS1**. TG 9 TS2 is often used for calls to a local regional reflector. If that is the case on your repeater, you could select TS2 instead.

    16) Digital Encryption should be **off**

    17) Encrypt Type – does not matter

    18) RX Group List – set to **None**

19) Work alone – not used

20) CH ranging **off**

21) APRS receive **off**

22) SMS forbid **off**

23) Data ack forbid **off**

24) DMR Mode **Repeater**

Associate the new channel with the new talk group, using the radio keypad.

I don't know why the add a channel process does not include allocating the talk group to the channel, but it doesn't. So back out of the menu tree and then, switch the channel knob on the radio until the display shows your new channel 'Local TG 9' on the zone, probably 'Zone1.'

Press **Menu > Talk Group > TG List** and scroll using the round **up-down** button to the **Local** talk group. Click **Select > Select Contact**. Now **Back** out of the menu tree.

Making a call

Now you should be able to make a call on your local repeater. If it is not working, the time slot or colour code may be incorrect, you can fix that by going back and editing the channel.

Finding the colour code and static talk groups.

The easiest way to find the colour code and static talk groups is to access the DMR network dashboard, but for this exercise, we are assuming that we don't have a computer.

We want the maximum possible information to be displayed when a call comes up on the repeater. Use, **Menu > Digi Moni > DigiMoni CC > Any CC** then **Back > Digi Moni ID > Any**.

If the channel B (lower) display is on, press and hold the P1 button to turn it off.

Activate the dual slot digital monitor on your radio by clicking the PF2 button twice. It is the lower of the two buttons below the PTT button on the side of the radio. It is indicated on the display with a red ◀)) icon. Or you can turn it on using **Menu > Digi Moni > DigiMoni Switch > Double Slot**. Leave 'Slot Hold' turned off.

Now sit back and wait for somebody to use the repeater. In the dual spot monitor mode, you should hear any call that is made. When a call comes up on the repeater, you should hear the caller and the display will show a name and callsign (or the DMR ID) of the caller, the talk group number and time slot (T) they used, and the colour code (CC) of the repeater. Armed with that information you can create another talk group and channel, with the new talk group, time slot, and colour code.

TYT MD-UV380 or MD-UV390 etc.

TIP: Don't stop! If you do, your changes will be lost, and you will have to start again.

The radio needs your DMR ID. Nothing will work without that.

Presuming that the 'Program Radio' and 'Edit Radio ID' features have been turned on, we will begin programming using the keys on the radio.

Menu > Utilities > Radio Info > My Number > Edit

Enter your DMR ID number. The red 'back' key backspaces. Then **Confirm** (Menu).

There is a 'Catch 22' situation here. You can only change the DMR ID on the radio if that feature has been enabled in the CPS program. It is under **General Setting**, check the **Edit Radio ID** checkbox. But if you are already using the CPS, it is much easier to program the radio from there. You can enter your **Name** and **DMR ID** on the same page. While you are there you should go to the **Menu Item** tab and tick the **Program Radio** checkbox. If they have not been tuned and you are unable to edit the DMR ID on the radio, skip the rest of this section and jump to 'Download the CPS configuration software' on page 22.

Now we need a talk group. The TYT radios label talk groups as 'contacts,' because they do not have a separate talk group list. But we are going to use the term talk groups because that's what the DMR networks and most DMR users call them.

At this stage, we don't know what talk groups are available on the repeater, but one thing is for sure, we will need the 'Local' talk group TG 9. Talk group 9 is used for talking over the repeater without the call extending to the wider network. On many DMR networks, it is also used for talking on 'reflectors' so, if the repeater is linked to a reflector, your TG 9 call may be heard over the entire region, state, or country.

Menu > Contacts > Contacts will display a list of the talk groups already on the radio. 'Local' or 'LCL' TG 9 may already be there. The radio will not save two talk groups with the same TG number. Assuming that TG 9 is not there already, use **Menu > Contacts > New Contact > Group Call**, enter a **9** and **Confirm**. Now enter **Local** using the keypad buttons. They cycle through lowercase, number and uppercase. Then click **Confirm**. On the **Select Tone** page, click **Confirm** again for **Tone Off**. That is all we need for a talk group.

Add a new channel using the radio keypad

Go to **Menu > Utilities > Program Radio >** go to **No.6 Add**, then select **Digital CH**.

The display will say Enter CH Name, Channel 1. Use the up ▲ arrow to backspace and enter a new name using the radio keypad buttons.

Multiple presses cycle through lowercase, a number, and uppercase. Press the **Menu** button to **Confirm**.

TIP: on the TYT radios all channel names must be unique. Even if they use the same talk group and are for a different repeater and in a different zone.

Use the up ▲ arrow to backspace and the keypad to enter the receiver frequency (the repeater output frequency). Press the **Menu** button to **Confirm**.

Use the up ▲ arrow to backspace and the keypad to enter the transmitter frequency (the repeater input frequency). Press the **Menu** button to **Confirm**.

Use the up ▲ arrow or down ▼ arrow the select the 'Local' talk group you created earlier. Press the **Menu** button to **Confirm**.

Use the up ▲ arrow or down ▼ arrow the select a receive group list. Press the **Menu** button to **Confirm**. Apparently, you cannot choose a 'None' option! Not essential.

Go to **Menu > Utilities > Program Radio**

No 7 Color Code. Use the up ▲ arrow or down ▼ arrow to select the **colour code**. Press the **Menu** button to **Confirm**. If you don't know the repeater's colour code, select **1** for now. It is the most commonly used colour code.

No 8. Time Slot. Use the up ▲ arrow or down ▼arrow then select the **time slot**. Press the **Menu** button to **Confirm**. The repeater time slot for TG 9 Local calls is usually **TS1**. TG 9 TS2 is often used for calls to a local regional reflector. If that is the case on your repeater, you could select TS2 instead.

This completes the 'add a new channel' menu structure. But there is no way to set the zone for your new channel.

Add your new channel to a zone

You have to add your new channel to a zone, or you will not be able to select it with the channel switch. Select **Menu > Zone > Zone List**. Select a **zone**. At this stage, there may only be one zone to choose from. Now select **Add CH > CH A**. Selecting CH A puts the channel into the A list (upper row). Now use the ▲ arrow or down ▼ arrow to find and select your new 'Local' channel. Now **Back** out of the menu tree.

Making a call

Now you should be able to make a call on your local repeater. If it is not working, the time slot or colour code may be incorrect, you can fix that by going back and editing the channel. You can find the correct colour code using the instructions in the next section.

Finding the colour code and static talk groups.

The easiest way to find the colour code and static talk groups for a repeater is to access the DMR network dashboard, but for this exercise, we are assuming that we don't have a computer.

Activate the digital monitor on your radio by clicking the PF2 button. It is the button below the PTT button on the side of the radio. The digital monitor is indicated on the display with a left-facing speaker icon ▶ on the top row of the display.

Now sit back and wait for somebody to use the repeater. You should hear any call that is made on the same time slot as the channel you made. When a call comes up on the repeater, you should hear the caller and the display will show the name and callsign (or the DMR ID) of the caller, the talk group they used, and the colour code (CC) of the repeater. If you get the display the time slot must be the same as your channel. Armed with that information you can create another talk group and channel, with the new talk group, time slot, and colour code.

If you see that the repeater is active with an RSSI indication and a green LED, but you cannot hear the repeater, chances are that your channel is on the other time slot. The TYT radios do not have a dual slot monitor. If you have a TS1 channel on the top (A) and a TS2 channel on the bottom (B), and you have the digital monitor on, you should hear any call on the repeater.

The display shown here may be only available after the firmware upgrade. I can't remember what was displayed before I did the upgrade, and I don't want to reload the old firmware.

'ID' is the DMR user ID of the calling station, 'CC' is the repeater colour code, TX Contact is the talk group, 'Type' is G for a group call, and 'Solt' is 2 for duplex frequencies and not shown for simplex frequencies.

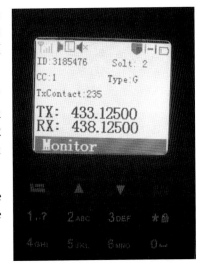

Including the frequencies is nice, but if they were wrong you would not receive the signal, so there would be no display.

TIP: By now you will realise that programming talk groups zones and channels from the keys on the radio is very frustrating. Time to get on with downloading and using the CPS configuration software. It really is the easiest way!

DOWNLOAD THE CPS CONFIGURATION SOFTWARE

Although you can program most radios from the radio keypad, you will quickly find that this is tedious, especially for digital channels where a lot of information has to be entered. It is certainly best to download and install the configuration software.

Each brand of radio, and often each radio model, has unique configuration software. Some radios are very similar or identical rebranded versions of radios supplied by another maker, so if you can't find the right software, you could try the software for a similar model. The box the radio came in or the supplied instruction sheet will probably include a web page URL for the product which should have a 'download' tab or area. I will step through the configuration process for my Radioddity handheld radio and similar models, and then the TYT radio and similar models. Your software will be similar, but probably not identical. Some features are not included on the cheaper radios. If a function is not available on your radio, just ignore the instruction and carry on.

The config software can read data from the radio. You can edit it and add new information then write the updated 'code plug' file back into the radio. You will need a 'programming cable' and again these are specific to the brand and sometimes the model of the radio. Try to buy a radio that includes the programming cable. Most do.

Virus detected! Is the CPS programming software safe to download? Windows is likely to throw up a blue unsafe application error screen when you attempt to install the CPS software. This is a 'false positive' caused by the Chinese software not having the approved software licence key. As far as I know, there has never been a problem with ignoring the alert and installing the software. I have downloaded several CPS programs. They all cause Windows to freak out, and they have all been 100% safe. Click 'More info,' then 'Run anyway.'

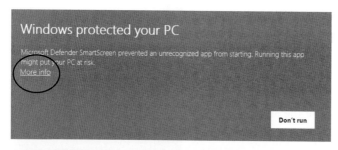

3ᴿᴰ PARTY CPS SOFTWARE

The CPS software for the Radioddity and AnyTone radios works very well, but the TYT CPS software for the MD-UV380 is more difficult to use. For some reason, TYT has a different application for the MD-UV390 even though they are identical other than the waterproof case. The TYT CPS software for the MD-380 and MD-390 is appalling with no ability it import or export .csv files. For that reason, I have included EditCP examples in the 'Programming with EditCP' chapter.

EditCP

Dale Farnsworth wrote an Open Source program called 'EditCP' for the TYT MD-380, MD-390, MD-2017, MD-UV380, MD-UV390, and MD-2017, the Alinco DJ-MD40, the Retevis RT3, RT3-G, RT3S, and the RT82 radios. It is available at https://www.farnsworth.org/dale/codeplug/editcp/. I had the same issue with Windows initially blocking it as I did with the TYT and Radioddity software.

EditCP can export all of the configuration data in one Excel spreadsheet. You can update the tabs by pasting in data from various sources, such as contacts from RadioID.net. You can also read a code plug from one radio and save it as a code plug for a different radio. Great if you happen to migrate from an MD-380 to a MD-UV380.

If you have a TYT MD-380, TYT MD-390, Retevis RT3 or Retevis RT8, you can use EditCP to update the firmware on the radio. [Use at your own risk].

Other 3ʳᵈ party software

There are others, but I had limited success with them. They include N0GSG Contact Manager which seems OK for older radios, DMR Codeplug editor by G6AMU which would not read any of my existing .csv or .rdt files but other than that seemed to work OK, and Chirp which does not support either of my DMR radios. They all look like good programs, and they may be fine for your radio.

THE CODE PLUG

A code plug is simply the information file that the PC configuration program downloads to your radio.

It's an old Motorola term dating back to the days when radio channel and configuration data was burned into a PROM (programmable read-only memory chip) that was plugged into the radio, i.e. software **code** that was **plug**ged in. This was to stop customers from changing the frequencies and other parameters. Only a qualified Service Technician was allowed to make changes. Some documentation refers to the code plug as a 'personality.' Even earlier than the PROM and EPROM versions, some radios required an external plug containing wire links, or diodes, that were used to make jumper settings or choose channels in the radio.

THE CONTACT LIST

The contact list contains the callsign name and other details of other registered DMR users. You can elect to add contacts yourself making a small contact list containing the callsign and details of people you call often. Or you can download a contact list from online sources. RadioID.net https://www.radioid.net/ maintains the register of all DMR users so it is a logical place to start. Once you are registered you will have access to the 'Contacts Generator' and the raw Excel data files. This service is not free. It costs $ 9.99 (USD) per year for unlimited use. An alternative is to use https://amateurradio.digital/ which costs $ 12 (USD) per year for unlimited use. Brandmeister has a free contact list download based on the users of specific BM talk groups at https://brandmeister.network/?page=contactsexport.

Step 1 is to select the contacts you want.

The RadioID website has an excellent video showing how to select contacts from your continent, country, region, state, U.S. County, or Canadian Province. You can also search for specific callsigns to make sure that your friends are included.

When you have selected some contacts, you can choose an 'email on difference' setting which will send an email when the list has grown by a preset number, 100, 500, 1000, or 10,000. Choose a low setting if your contact list is small. If your contact list is large, for example worldwide or all of the USA, select one of the higher options, or you will be getting emails every day or two.

Step 2 is to define the format that you need to download for your specific radio.

Your specific radio might not be listed, but you can probably select a model from the same manufacturer. I used the download for the AnyTone AT-D878UV since it is compatible with my Radioddity radio, which is not listed. For the TYT MD-UV380, I selected the file for the TYT MD-UV390 (GPS). There is an option to create a custom list for your radio. Or, if you get really stuck, you can edit the downloaded file using a spreadsheet program to move the data into the correct spreadsheet columns.

Step 3 is to download the file.

You need the 'comma delimited' .csv file not the .xlsx Excel file. If you check the TG check box, the file will also include the 1606 Brandmeister talk groups. Select this option if you have a TYT radio and you will be using the Brandmeister network. Large downloads may take a minute or two. If you have not registered and paid for the Contacts Generator service, you will be invited to do so before the file can be downloaded.

The file will be different for various radio brands. The AnyTone/Radioddity version has columns for the Radio ID, Callsign, Name, City, State, Country, Remarks, Call Type, and Call Alert. The TYT version has; the Contact/talk group name, Call Type, Call-ID (TG or contact number) and Call Receive Tone.

Save the downloaded .csv file in a convenient location on your PC. I made a 'My Documents\DMR\CPS' directory. You have to retrieve the file using the CPS programming software so that it can be sent to the radio.

TIP: The 'Digital contact list' on a Radioddity/AnyTone radio, or the 'CSV contact list' on a TYT radio with the firmware upgrade, is only used to display information about DMR stations that you hear. Not for calling anyone using the Private Call mode.

TALK GROUPS

A DMR talk group is a way of grouping many stations together. If you transmit on a talk group, every radio that has the same talk group selected can hear you and has the opportunity to respond to your call. They can be using any DMR repeater or hotspot that is on the same network (or a linked one) and the same talk group (or a linked one) worldwide. If your radio is listening to the same repeater but a different talk group is selected, you will not be able to hear the call, even though the repeater is transmitting the data. You can add talk groups to your 'receive group call list' if you want to receive them but not transmit. Most radios have a 'promiscuous' mode or 'digital monitor' which let you hear all of the DMR transmissions from the repeater.

Talk groups can exist for many reasons. There are talk groups for countries, states, provinces, counties, regions, cities, and special interest groups. Talk groups are specific to the individual DMR networks, but they all generally follow a similar numbering scheme based on the MCC (mobile country codes). When two networks use the same talk group (TG) number it sometimes indicates that the talk groups are linked between networks. Each network publishes a list of its talk groups.

Be careful to identify talk groups from different networks and use zones to prevent mix-ups. For example, if your repeater is on the DMR+ network, but your hotspot is on the Brandmeister network, create a hotspot zone and a separate repeater zone.

Talk group links

DMR+ talk groups are at, https://www.pistar.uk/dmr_dmr+_talk_groups.php

DMR+ reflectors are at, https://www.pistar.uk/dmr_dmr+_reflectors.php

TGIF talk groups are at, https://www.pistar.uk/dmr_tgif_talk_groups.php

The K3NYJ blog has TGIF talk groups at http://k3nyj.blogspot.com/2020/06/talk group-list-tgif-network.html

Brandmeister talk groups are at, https://www.pistar.uk/dmr_bm_talk_groups.php

UK talk groups are at, https://www.dmr-uk.net/index.php/layout/ and

http://www.g0hwc.com/downloads/Phoenix%20Network%20Talk%20Groups%20v1.9.pdf

There are a few more links on the 'Internet links' and 'great videos' pages.

Special talk groups

TG 9 TS1 is used for calls that will only be transmitted from your local repeater. TG 9 TS2 is often used for calls to an already connected reflector.

TG 8 TS1 is typically for your local region. The region is programmed for the repeater and cannot be changed by the repeater users.

TIP: If you add DMR+ as a second network on your hotspot, you will use TG 9 for Brandmeister and TG 8 will be translated to TG 9 for the DMR+ reflectors.

TG 99 TS1 is usually allocated to Simplex use. CC1 (colour code 1), Admit Criteria = Always, In Call Criteria = TX or Always. Frequencies will vary from country to country. Possibly, UHF 441.000, 446.500, 446.075, or 433.450 MHz and VHF 145.790 or 145.510 MHz. We use 432.750 MHz in New Zealand. Australia uses 439.200 and 144.800 for simplex. 439.150, 439.125, 145.325, 145.350, 145.375, 147.525 and 147.550 for hotspots and 144.750 for high power and high altitude hotspots.

TG 98 on the Brandmeister network is available for you to make test calls.

TG 4000 is used to disconnect dynamic talk groups, except on Phoenix UK where TG 400 is used instead. TG 400 will only disconnect a user access talk group that you have initiated. TG 4000 is not used on the TGIF network.

TG 9990 on most networks is a 'Parrot' or 'Echo' which will send back your transmission so you can hear what it sounds like. Usually, TS1 is for hotspot users and TS2 is for repeater users.

Many networks do not allow APRS because the beacons hog the bandwidth, but some have a talk group especially for APRS. Try 5057 on Phoenix UK, 234999 UK and 310999 USA on Brandmeister.

Static and dynamic talk groups

Most repeaters are linked to static talk groups. When nobody is using the repeater, it stays linked to the static talk group(s) and will transmit any calls that are on the talk group(s). The DMR+ static talk groups are listed on the IPSC2 dashboard in the 5th column for TS1 and the 8th column for TS2. You can switch to a 'dynamic' talk group by 'keying up' and transmitting on it. After you finish with the repeater, it will drop back to the static talk group after a waiting period of (usually) 15 minutes.

The Brandmeister network also has auto-static talk groups which are used instead of dynamic talk groups when you are using a simplex hotspot. Auto-static talk groups stay connected until you key up a different talk group. More on page 161.

The TGIF network makes any call or 'key up' to a talk group auto-static until you transmit to a different talk group or change the talk group on the TGIF dashboard.

UK calling channels

In the UK, TG 1, TG 2, TG 13, and TG 235 are "calling" channels. Once you have established a contact, please choose a User Activated (UA) talk group and arrange with the other station to move your conversation there, to free up the calling channel for other users. User Activated (UA) talk groups include TG 119 (Worldwide UA1), TG 129 (Worldwide UA2), TG 113 (English Worldwide UA1), TG 123 (English Worldwide UA2) or TG's 80, 81, 82, 83, & 84 (UK Wide UA1-UA5). I am not sure how often this etiquette is followed. User activated is just another way of saying dynamic.

U.S. calling channels

TAC, 'talk around channels,' shouldn't be used as primary calling channels. The U.S. TAC channels (310 - 319) cannot be added as static talk groups and will not become auto-static. If you want a static primary calling channel, try one like U.S. Wide 3100.

DMR NETWORKS

There are several large and many regional or special interest DMR networks. Each has its own set of talk groups (TG). Sometimes two networks will use the same TG number for different groups. But most often the numbering, at least for the most popular talk groups does not overlap. Where the big networks use the same TG number it usually means they cover the same geographical region. It might indicate that the talk groups are linked, and you can use either network to reach the same people. Or they could be completely unrelated.

Worldwide networks

- Brandmeister is the largest DMR network, popular in the USA, 1606 TGs

- DMR+ is the original Tier II network, most popular in Europe, 231 TGs

- TGIF, the Thank God it's Friday, informal & special interest network, 727 TGs

- DMR-MARC is the original DMR network, 73 TGs. Only connects Motorola MotoTRBO repeaters via C-Bridge connections, no hotspots. DMR-MARC is linked to the DMR+ network.

Regional and national networks

There are many regional networks. They usually offer a range of local talk groups that are only used within the network, and they are usually linked via IPSC2 servers to DMR+ to provide their users with worldwide access.

- CAN-TRBO – Canadian network, 29 TGs

- ZL-TRBO – New Zealand AREC and general ZL network, 14 repeaters

- VK DMR – Australian network

- FreeDMR UK – England, Ireland, Scotland, & Wales network, 26 TGs

- Phoenix UK – England, Ireland, Scotland, & Wales network, 15 TGs

- Other UK networks include the Northern DMR Cluster (NDC), SALOP Cluster, South West Cluster (SWC), DV Scotland, and FreeSTAR.

- As you would expect, there are hundreds of US regional networks. Many are local networks or repeaters connected via c-Bridge to the DMR-MARC network. DCI TRBO, Lonestar, DMRX, K4USD, Crossroads, Chicagoland, Rocky Mountain Ham, and Bronx TRBO are just a few. PNWDigital.Net (Pacific North West) connects 55 MotoTRBO repeaters.

Some talk groups are permanently interconnected to talk groups on other networks, and some networks are linked to other networks. For example, the DMR-MARC and DMR+ networks are permanently linked. A few repeaters offer static talk groups on two different networks and many servers connect DMR systems to talk groups, 'reflectors,' or 'rooms' on AllStar, Echolink, D-Star, or System Fusion networks.

For example, the ZL-TRBO network is on a MotoTRBO platform (DMR-MARC network) providing two national talk groups, over 14 repeaters. It is connected to the DMR+ network via a c-Bridge link to provide international access and dynamic talk groups. You can reach most New Zealand DMR users via DMR+ TG 530. Brandmeister TG 530 is not linked and will only reach ZL hams with Brandmeister hotspots.

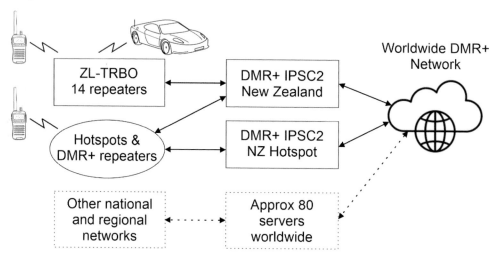

Figure 7: A typical local network connected to a DMR+ network

Brandmeister is the biggest both in terms of talk groups and connected repeaters and hotspots. DMR+ is not widespread in the U.S.

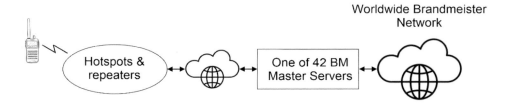

Figure 8: A Brandmeister network connection

DMR REFLECTORS

DMR reflectors work the same way as they do on D-STAR or IRLP. When a station transmits on a talk group that is connected to a reflector, their signal is transmitted by all other nodes connected to the same reflector.

The difference between a static talk group and a reflector is that you have to specifically link and unlink a reflector. This means that only the repeaters and hotspots that are connected to that reflector are tied up during transmissions, not hundreds of repeaters on worldwide talk groups. Reflectors usually time out after 5 minutes of inactivity, compared with dynamic talk groups which typically time out are after 15 minutes of inactivity and static talk groups which don't time out at all.

You can see 'Reflector' connections on the Remap tab of the IPSC2 dashboard for your network. See the sections on IPSC2 and Brandmeister dashboards on pages **Error! Bookmark not defined.** and 137 respectively.

Reflectors are numbered from 4000. There is a list of the ones that can be accessed via DMR+ at https://www.pistar.uk/dmr_dmr+_reflectors.php. Some reflectors are links to D-Star reflectors or YSF Rooms. For example, reflector 4444 is DCS001 and 4454 is XLX600E. Reflector 4409 is the System Fusion Wires-X CQ-UK room.

If the reflector is not associated with a static talk group,

1. Transmit on TG 5000 TS2 to check the repeater's link status

2. Key up on TG 4xxx TS2 for a few seconds to connect to a reflector

3. Transmit on TG 9 TS2 to talk to a station or make a call

4. At the end of the QSO transmit on TG 4000 TS2 to unlink the reflector

Some repeaters have static talk groups already associated with reflectors. The New Zealand repeaters have static talk groups on TG 153 TS1 for the South Pacific (SPAC) reflector 4851, and TG 530 TS2 for the New Zealand national reflector 4850. You can connect to the reflector and talk on TG 9 or connect to the talk group and talk on the talk group number.

USING A DOWNLOADED CODE PLUG

The next sections step you through programming your radio. You use the CPS (customer programming software) and optionally a spreadsheet program or the radio keypad to create a code plug, which you then download to your radio.

It is possible to download a completed code plug .rdt file or use one that your local DMR group or a friend has created. Open it in your CPS software and then download it to the radio. This can save you a lot of work, but you must be careful that it contains the information that you need. I downloaded a New Zealand ZL code plug for the AnyTone AT-D878UV, and it works fine except that it did not have any 5 MHz repeater offsets programmed.

At the very least, you **MUST** change the DMR ID to your ID number before sending the file to the radio. Either way, the next chapter will be useful if you want to make any changes going forward.

READING AND WRITING DATA TO THE RADIO

Reading data from and writing data to the radio is very much the same as programming an FM handheld using PC software. To make any changes you start the CPS (customer programming software), connect the radio, load the last configuration file you saved on your hard drive, or upload the current configuration from the radio, edit the configuration, save a backup on the PC, and then send the updated file back to the radio. It only takes a few minutes. Most CPS software only runs on Windows.

Programming cable

Most radios come with a 'programming cable.' If not, you will have to buy one specifically for the radio model. Typically, the radio end is a 'mic & speaker' phone plug, but it could be a 'mini USB' connector or a propriety microphone connector in the case of Tait, Hytera, and Motorola radios. The PC end will almost certainly be a USB Type-A connector. Some cables have a built-in serial data interface, others are just a straight cable. If the radio requires a cable with a data interface, you must not use a straight cable version and vice versa. Without the appropriate voltage conversion, your radio might be damaged. My Radioddity radio came with a straight cable that plugs into the mic/speaker jacks on the radio. But the TYT cable has a USB driver in it.

TIP: Be very careful not to accidentally hit the PTT (Push to Talk) button while you have the USB cable plugged into the radio, or while it is sitting in the charger. RF energy can be coupled into your computer or the computer part of the radio causing lockups and possibly failure of the radio. It is very easy to push the PTT while you are plugging in the programming cable. I try to get into the habit of turning the radio off while I plug it in. Or you could make sure to only plug in the USB end of the cable after you have plugged in the radio end.

Connect the cable

Plug the cable into the radio and any USB port on the PC and turn the radio on. Windows 10 will recognise the device and it is unlikely that you will need to download a device driver.

Set the COM port

The TYT software detects the COM port automatically, so you can skip this section.

For the AnyTone, Radioddity, and similar radios. Start the CPS software that you downloaded earlier, (page 22).

Upload and download data from and to an AnyTone/Radioddity radio

Click the 5th icon. The one that looks like two linked boxes or click **Set** then **Set COM**. Select the Com Port for the radio. If you don't know which COM Port to use, click the icon  and then unplug the USB cable from the computer. The COM port that disappears from the list and reappears when you plug the cable back in is the one to use. Do not transmit while the cable is plugged into the radio! It is best to put the radio down on the table or into its charging stand.

To upload the existing configuration from the radio, click the 4th icon. It's the one that shows an arrow pointing at a computer. Or click **Program** then **Read From Radio**.

To download the updated 'code plug' configuration back to the radio, click the 6th icon. It's the one that shows an arrow pointing at a handheld radio. Or click **Program** then **Write to Radio**. A popup window will say that the computer is reading data from the radio, and a warning on the radio screen tells you not to turn the radio off or pull out the cable.

Upload and download data from and to a TYT MD-UV380/390 radio

To upload the existing configuration from the radio using the TYT CPS, click the 4th icon. It's the one that shows an arrow pointing away from the handheld radio. Or click **Program** then **Read data**.

To download the updated 'code plug' configuration back to the radio, click the 5th icon. It's the one that shows a red arrow pointing towards a handheld radio. Or click **Program** then **Write data**.

Upload and data using EditCP

On EditCP, click **Radio > Read Codeplug from radio**. The COM port is detected automatically when you plug in the radio and USB programming cable. To write the edited code plug back to the radio, click **Radio > Write Codeplug to radio**.

Using DMR technology

This chapter contains some facts that will help you understand the dynamics of using the worldwide DMR networks before we get into programming the radio. It might answer some of the questions that have been worrying you.

Of course, you can just program a few local talk groups, and use your local repeater to talk to 'the locals' or across the state, province, or county. One talk group, one channel, one zone, and you are on the air!

1. To hear a conversation on your local repeater or your hotspot, you need at least one **channel** configured for the correct frequencies. Just like you do for an FM repeater.

2. Normally the channel needs to be associated with the **talk group** that is being transmitted. It also needs to be set for the **colour code** that the repeater uses and the **time slot** that the repeater is using for that particular talk group.

There are two exceptions to this rule. If you turn on the digital monitor (promiscuous) mode, you will be able to hear all the traffic on the same time slot as your selected channel, even if you don't have that talk group loaded into the radio. The AnyTone/BTECH/Radioddity radios have a dual-slot monitor mode. If that is selected, you will be able to hear all traffic on the repeater or hotspot irrespective of the time slot that is being transmitted. This can be a useful feature if you are having configuration problems and you want to find out what talk groups are being used and which time slot and colour code they are using.

The second method is to activate a 'receive group' list. As well as the talk group that is associated with the channel, you will hear all the talk groups that you have placed in the selected receive group list. Provided that they are being transmitted by the repeater or hotspot and they are being transmitted on the same time slot as the selected channel. Receive groups can cause a whole lot of confusion with you not being able to transmit back to a station that you can hear. My preference is to establish two monitor channels, one for each time slot, and attach them to two receive groups containing your preferred talk groups. For all of the other channels, I set the receive group list to 'none.' That way I can switch to the monitor channel and hear everything. When I want to make a call, I switch to the correct talk group and can only hear the traffic on that single talk group. Which is what you want when you are talking with somebody.

You could elect to not use receive groups at all and use the digital monitor when you want to listen to everything on the time slot.

The main difference is that you can pick the talk groups in a receive group list instead of getting everything that the repeater transmits.

3. This process is repeated every time you set up a new channel.

TALK GROUP ➜ CHANNEL ➜ ZONE

    a. Identify a talk group you want to use. It must be on the DMR network that the repeater or hotspot is connected to. My hotspot is connected to both Brandmeister and DMR+ which adds a layer of complexity. But I will get to that later, in the Pi-Star chapters.

    b. If the talk group is not already in your radio, it needs to be added.

    c. Then you create a channel that matches the repeater frequencies, colour code and time slot, for the new talk group.

    d. You won't be able to select the new channel with the channel switch on the radio unless you add the channel to a zone.

    e. You might optionally want to add the new talk group to the receive group list for the time slot, or the new channel to the scan list.

4. Repeaters and duplex hotspots use two time slots (TS1 and TS2). Simplex hotspots use one time slot (TS2).

5. If you want to transmit on a talk group. It must be in the radio, and it must be associated with a channel that matches the repeater frequencies, colour code and time slot. Finally, the channel must be in a zone.

6. Yep! There is an exception to this rule. Some people use this mode of operation all the time and don't bother loading a lot of channels and zones.

    a. Select any channel that is on the time slot and frequencies you want to transmit on.

    b. Turn on the digital monitor and identify a talk group that you would like to use, (optional).

    c. Select any talk group that is on your radio using the menu and keypad buttons on the radio. Or use 'manual dial' to enter any talk group number or private DMR ID number.

    d. Now you will transmit on the selected talk group even though the displayed channel information has not changed.

    e. I am not a big fan of this method. For one thing, the radio display does not show the talk group you are transmitting on. Also, the radio will revert to the displayed channel after a while. I prefer to know what will happen when I press the PTT.

7. Talk groups act in different ways (repeater access).

    a. The repeater owner will specify what talk groups you can use on TS1 and what they would prefer to be on TS2.

    b. The network server owner and the repeater owner will make some talk groups 'static.' Static talk groups usually include your regional talk groups and a few popular talk groups. They may exclude extremely popular talk groups like Brandmeister TG 91 on the grounds that it is so busy that it is difficult to get in and make a call on a different talk group. Some busy talk groups cannot be made static. All calls on a static talk group will be transmitted by the repeater unless it is already transmitting another talk group on that time slot. This is why repeater owners will usually specify that local calls be on a different time slot.

    c. On a repeater, if you 'key up' (press the PTT) with your radio set to a channel that is on the correct network, but is not static, it will direct the traffic on that talk group to your repeater for a period, usually limited by 15 minutes of inactivity, and then will drop it again. It is called a 'dynamic' talk group.

8. Talk groups act in different ways (hotspot access).

    a. If you own a hotspot you get to make some of the rules, depending on the DMR network you are using and the type of hotspot. On Brandmeister, you can select your choice of static talk groups on the international Brandmeister dashboard. If you have a duplex hotspot, you can make some talk groups static on TS1 and some on TS2. If you have a simplex hotspot, you can only allocate static talk groups to TS2.

        TIP: My duplex hotspot is connected to two DMR networks, I use, one time slot for network one, and the other for network two. That way I know when a call comes up on TS1 that it is a Brandmeister talk group, and if it comes up on TS2 it is a DMR+ talk group. You don't have to do this, But I like it.

    b. On the DMR+ network and all networks connected to IPSC2 servers; the static talk groups are set by the server owner. You can nominate which of those you want to hear (as static) in your hotspot configuration software. You may only want to hear one or two of them.

    c. The hotspot configuration software will be Pi-Star for most hotspots. If you own a duplex hotspot, you can force the time slot to change.

For example, I have all of my DMR+ activity on TS2. But the DMR+ server wants some of them to be on TS1. You can change the settings in Pi-Star to achieve that.

d. A duplex hotspot can also create dynamic talk groups. If you 'key up' (press the PTT) with your radio set on a channel that is on the correct network, but is not static, the network will direct the traffic on that talk group to your hotspot for a period, and it will drop out (usually) after 15 minutes of inactivity.

e. On DMR+ and other networks connected to the DMR+ network via IPSC2 servers, a simplex hotspot acts the same. If you 'key up' a non-static channel, the network will direct the traffic on that talk group to your hotspot, and it will drop out after 15 minutes of inactivity.

On Brandmeister, when you 'key up' a non-static talk group via your simplex hotspot the system creates an **auto-static** connection. It works the same except that it never drops out. It remains static until you key up a different talk group. When you do that, the new talk group becomes auto-static and the old one becomes dynamic and should time out after 15 minutes. Hearing or seeing both talk groups on your hotspot can be distracting, so it is best to key up the 'disconnect' TG 4000 talk group to disconnect the auto-static talk group, before keying up the new talk group.

9. The DMR+ network supports **reflectors**. Like talk groups, they allow you to talk with other hams using other repeaters and hotspots. They work in the same way as D-Star reflectors and System Fusion 'Rooms' and many of them support multiple digital voice modes. You can connect to a reflector and chat with people using D-Star or System Fusion radios. The Brandmeister network no longer uses reflectors, but some talk groups are interconnected with D-Star reflectors and/or System Fusion 'Rooms.'

The talk groups for reflectors are in the range between 4000 and 5000. You connect to one by keying up its talk group number, in the usual way. But after it is connected, you talk on TG 9 (Local). You disconnect in the usual way, by keying up TG 4000. TG 5000 responds with a link status report.

On a hotspot that is connected to two networks, it is common to transmit on TG 8 or another single number TG and the Pi-Star software will redirect it to TG 9 on the second (usually DMR+) network.

"What you see is NOT what you get." Talk groups are not the same on different networks. TG 530 on Brandmeister is 'New Zealand.' TG 530 on DMR+ is called 'ZL.' It is also a New Zealand national talk group, but they are not linked.

The 14 national network repeaters in New Zealand are on the ZL-TRBO network, which is linked to DMR+ TG 530 via a c-Bridge link to an IPSC2 server. I have to have access to a repeater or have DMR+ on my hotspot if I want to participate in the ZL chat and the weekly digital modes net.

The Brandmeister TG 530 talk group is mostly used by DX stations wanting to work ZL stations.

10. DMR works in layers. You don't have to use them all.

   a. You can talk directly to another DMR radio using digital simplex TG 99, or a private ID call.

   b. You can talk to a local ham over a local repeater using TG 9 TS1. Often TG 9 TS2 will be linked to other local or regional repeaters, so you may be able to reach stations outside of the coverage of your local repeater.

   c. You can talk to a ham who is on the same network, using a local network talk group. Perhaps over Phoenix UK, or the Northern DMR Cluster.

   d. You can talk to a ham who is on the same worldwide network. For example, any DMR+ talk group, or over the Brandmeister, or TGIF network.

   e. You can talk to a ham who is using a different digital voice technology such as Yaesu System Fusion, P25, or D-Star, via a DMR+ reflector or a linked talk group.

   f. In a few cases, you can talk to a ham using a different DMR network. Brandmeister does link to some DMR+ talk groups. If you have a hotspot you can connect to up to four networks. Or you can have two hotspots.

11. Bridges, hubs, and clusters. A bridge is usually a link to a talk group using different digital voice technology. Such as a DMR+ reflector, a D-Star reflector, a System Fusion Wires-X room, or a P25 talk group.

Hubs link different DMR or other digital voice systems using Digital Voice over IP Multimode Interlinked System (DVMIS). It provides connections to Allstar, D-Star, HB-link, EchoLink, P25, YSF, or Wires-X.

Clusters are groups of repeaters that have a single link into the DMR network rather than each repeater having an individual link.

A c-Bridge is a networking protocol used to join a repeater or a group of repeaters into a DMR network. c-Bridge connections were originally used to connect MotoTRBO repeaters into the DMR-MARC network, but now they could be linking to another small network, a regional network, or the Brandmeister network instead.

The Brandmeister network

I believe that the Brandmeister network is the easiest to use and it has the most talk groups. It is more intuitive and the Brandmeister dashboard has a heap of live information. Brandmeister is the dominant DMR network in the USA.

It is very easy to configure your own choice of static talk groups on time slot 2, and time slot 1 if you are using a duplex hotspot. Unlike the other networks, they are added on the main dashboard, not on your hotspot's Pi-Star dashboard. If things are quiet, and I am not sure that everything is working, I can pop onto the website and set TG 91 to one of my time slots. There will be regular calls every few minutes. Keying up any talk group that is not on your static list will set it to dynamic and you will hear it for 15 minutes from the last time you transmit on the channel.

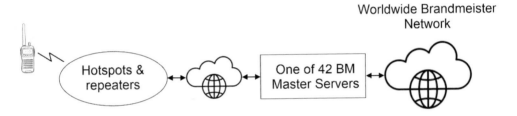

Figure 9: The Brandmeister network

The Hoseline and Player features are great, more about those later. You can also look at the 'last heard' list and filter it to see if a particular talk group is active or search for a particular name or callsign.

I set up monitor channels so that I can hear any of my static talk groups that become active. If I hear someone I want to call, or I get sick of the chatter, I switch to another Brandmeister channel where I will only be able to hear that individual talk group. If you are using both time slots you will need two monitor channels because the receive groups only work on the time slot associated with the channel you have selected.

For me, the only downside of the Brandmeister network is that I can't use it to talk to users of the New Zealand ZL-TRBO repeaters. Our local digital voice net is held on the ZL-TRBO network. I can only participate via a repeater, or via my hotspot using DMR+ TG 530. It is the main reason that I added DMR+ to my hotspot.

Many other local DMR networks, especially in Europe, are also connected to the world via the DMR+ network. If you want to talk to hams that are using a local network such as, Phoenix or DV Scotland you either need to access a repeater on that network or the equivalent talk group or reflector on the DMR+ network.

The DMR+ network

DMR+ is the dominant DMR network in Europe. It has a different design philosophy from Brandmeister. As well as supporting direct hotspot and repeater connections, the network ties together many regional and national DMR networks. It is an addition to the local DMR networks that allows worldwide contacts, which is why it is called DMR plus.

When you make a local or regional call, the server routes the traffic through the local DMR network. If you select a talk group that is outside the local network, it will be routed through the DMR+ network to the IPSC2 server in that country or region. Or it might be a worldwide talk group carried on all DMR+ servers. The local network may have 'local access only' talk groups that are not available on the wider DMR+ network.

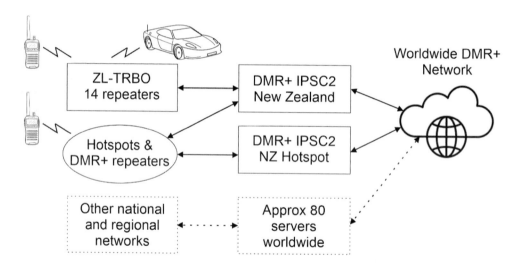

Figure 10: The DMR+ network

Unlike the Brandmeister and TGIF networks, there is no central dashboard for the whole network, so it is difficult to work out how busy it is. Each DMR+ IPSC2 server has a separate dashboard. There are currently 80 DMR+ servers. Each one supports dozens of repeaters, and often hundreds of hotspots, plus links to other DMR networks, reflectors, and different digital voice services such as Yaesu System Fusion rooms, and D-Star reflectors.

DMR+ supports normal DMR talk groups and DMR reflectors which are used in a different way. I am not sure why the DMR+ team feel the need for reflectors, but I am sure they have a good reason. It is probably to support cross digital mode voice traffic.

You can see that the network diagram is significantly more complicated than the Brandmeister drawing. This is a little unfair because the Brandmeister network also supports the connection of smaller Statewide or regional networks, often linked using the C-Bridge protocol. The point to note though is that DMR+ works this way all the time. The network drawing shows the New Zealand situation, but you could change the names and it would look much the same in England or other European countries. Regional or nationwide DMR networks connected via IPSC2 to the DMR+ network.

The drawing shows a box representing the ZL-TRBO network of 14 repeaters which provide nearly nationwide coverage. We can use the ZL-TRBO repeaters to talk to users on any of the other repeaters. ZL-TRBO links into the DMR+ network via a c-Bridge link to the 'New Zealand' IPSC2 server, and from there to the world via DMR+ talk groups. The IPSC2 server allows connections from repeaters that are not part of the ZL-TRBO network and from hotspots. There is a second DMR+ server in New Zealand, called 'NZ Hotspot.' It allows connections from repeaters and hotspots. 'NZ Hotspot' is not directly linked to ZL-TRBO, but you can make a call on "ZL" DMR+ TG 530 and it will be broadcast by the ZL-TRBO repeaters.

The way that static talk groups are arranged on the DMR+ and connected networks is different as well. On Brandmeister, you can choose the static talk groups for your hotspot. On DMR+ you can only choose from the selection that the server owner has made available. If you have an ongoing need, you can probably ask the server owner to add another talk group. The server owner will also decide which time slot may be used for calls over the local network and which may be used for overseas DMR+ talk groups. If you are using a repeater, the static talk groups are pre-set by the repeater owner and will be a subset of the talk groups that are available on the server.

You can use static talk groups anytime. They are always broadcast by the repeater or hotspot. Keying up a talk group that is not static will make it dynamic until you have stopped transmitting on the channel for 15 minutes. The rules are a little different for simplex hotspots, but I will cover that later.

There can be latency (delay) when you access a local network via a DMR+ talk group. Using the local repeater is faster. If you use two radios to monitor both options, the delay on the DMR+ link is quite noticeable.

DMR+ talk groups work the same as talk groups on any other DMR network. You key up a channel and it remains linked permanently or for about 15 minutes. Making a group call, 'keys up' all other repeaters and hotspots that are also connected to that talk group. Reflectors work differently. You key the channel to establish a link to the talk group, but after that you transmit and receive the reflector, using Local TG 9 (private call). You will be heard by any station connected to the same reflector and you will hear any calls on the reflector.

The TGIF network

The TGIF (Thank God it's Friday) network has a more laid back feel. It caters to special interest groups and those who want longer conversations. A lot of TGIF talk groups are linked to reflectors, talk groups, and rooms for other digital voice modes. The network was started in October 2018 by Robert K4WZV and Mitch EA7KDO with five users and one talk group TG 31665 running on a Raspberry Pi. Now it has 798 talk groups. The current 'Prime TGIF' version is still under development. It is working well but there are still some features to be added. For example, the network does not support private ID calls. There are talk groups for many radio clubs, Flying Hams, General Motors ARC, Computer Geeks of America, HF Portable Operation, Motor Coach Operators, ELECRAFT all products, worldwide DX, Secret Squirrel Radio Network, Ghost Spirits Demons & Paranormal, Wheel Tappers and Shunters Club, DoDropIn, and my personal favourite, the Drink Beer Get Drunk & Talk DMR talk group.

It is preferable to register on the https://tgif.network/index.php website if you want to use the TGIF network. You will be issued an automatically generated hotspot password. The generic password still works, but not for all talk groups. Registration is free, quick, and easy.

The website has the usual help screens and links to some helpful videos. You can list the talk groups that are currently active and see the 'last heard' talk groups. This can be filtered by entering items into the search box. For example, entering a partial callsign, a talk group number, or a first name. The pause icon stops the chaos so you can see what is happening. The 'active talk groups' page includes the callsign of the station using the talk group. A blue icon with a callsign and first name indicates that the station is linked to the talk group. Click on the link to show the station details. A grey icon with no name indicates that there is no station information. A green icon indicates that there is voice traffic, or the station is 'keyed up' on the talk group.

STATIC TALK GROUPS

Static talk groups work a little differently on the TGIF network. Keying up a TGIF talk group on either time slot makes it auto-static. The time slot will stay allocated to your hotspot until you key up a channel with a different talk group.

You can also set the talk groups that will be sent to your hotspot in the Selfcare page on the TGIF dashboard. The best way to know what talk groups are active is to view the 'last heard' and 'active talk groups' pages on the TGIF website (dashboard).

TIP: the TGIF network does not use the TG 4000 disconnect. Just key up a different talk group and the network will swing over to that. This can make it difficult to change talk groups. You have to wait for a gap in the conversation and press the PTT to move to a different talk group. Sometimes there are no gaps in the conversation!

Programming your Radioddity or AnyTone radio

Programming options vary according to what radio you have, but the basics will be the same. This chapter covers programming instructions for the AnyTone AT-D878UV and the Radioddity GD-AT10G. It should also be good for other radios based on the D878U template. The BTECH 6X2 is based on the D878U template, with changes designed especially for amateur radio operation. And the CPS software for the AnyTone D878UV and the Alinco DJ-MD5T is very similar, so these instructions should broadly apply with some minor variations.

The CPS software for the Radioddity radio is excellent. There is context-sensitive help at the bottom of every setup screen, and you can upload files in .csv format for any of the tables, (contacts, talk groups, zones, channels, APRS, hot keys, scan lists etc.).

The following chapter covers the TYT MD-UV380 and the TYT MD-UV380 and similar models, which have quite a different CPS program.

Start the radio programming software that you downloaded from the manufacturer's website earlier. Most manufactures and websites refer to CPS (Customer Programming Software) so that's what I will do as well. It is just another name for the radio configuration software, although how you "program a customer" is unclear.

Start your radio and connect the programming cable between it and the computer. Plug the radio end in before you plug the computer end in. The computer should recognise the device with the usual 'ping.' There is usually no need to load a device driver. Do not transmit while the cable is plugged in!

It is good practice to upload the present radio condition to the PC before you make any changes and save it to your PC as a backup file. Just in case you want to backtrack. If you don't, you will lose any changes that you made via the keys on the radio when you send a new file to the radio. It ensures that you are making changes to the current radio configuration, not some earlier iteration.

AFTER MAKING CHANGES IN THE CPS PROGRAM

Of course, you have to save the edited 'code plug' back to your radio before the changes you have made in the CPS will take effect. I always save a backup copy of the new configuration on my PC as well. Click the **disk icon** or use **File > Save** or **File > Save As**. The default file is called **'new.rdt'**. I saved my file as **GD AT10G.rdt**. You could save several different versions if you wanted to be able to return the radio to different configurations. The CPS program will automatically load the last saved file the next time you run the software.

NEXT STEPS

This is what we are going to do next.

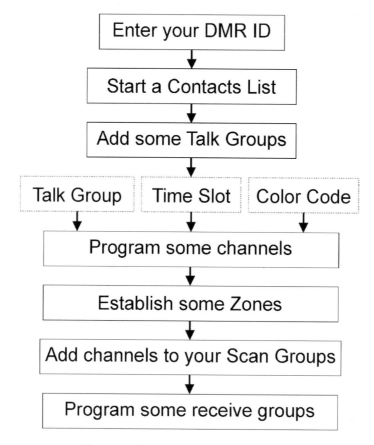

Figure 11: DMR programming stages

ENTER YOUR DMR ID

This only needs to be done once unless you replace the code plug with one from another source. If you do that, you **must** change the DMR ID before using the radio.

TIP: My radio can store 250 DMR IDs. Usually, you only need one. However, if a licenced family member or a club member will be using the radio, or the radio is used by several people who are part of an emergency response group or your knitting circle, you can enter their DMR ID numbers. If you will be using the radio on a commercial DMR network as well as for amateur radio, you can enter your commercial DMR ID and your amateur radio DMR ID.

On the Radioddity CPS, the DMR ID is at, Digital > Radio ID List. Double click line 1 and enter your DMR ID and a name or identifier for the radio. The radio name is usually your callsign - name, or just your callsign. It doesn't matter.

IT'S ALL TOO TECHNICAL WHAT SHOULD I DO?

All this uploading and downloading files and fiddling with Excel spreadsheets seems way too difficult. Is there an easier way? Sadly, the answer is "no." Unfortunately, DMR was never meant for amateur radio. The configuration of commercial radios is done by the "Technical Boffins." But in the amateur radio world, we have to do it ourselves. That's why we are interested in amateur radio, right? Although it seems challenging, you will soon find that it is not too bad and learn something about spreadsheets along the way.

Once all the channels, talk groups, and zones have been loaded, using the radio is relatively easy. You only need to set up the radio once. It is the same experience as when you had to add a heap of repeaters to your FM handheld or mobile.

It is possible to program most DMR handheld radios using the buttons on the radio keypad. But you have to go through the same steps, creating talk groups, channels, and zones, and it is extremely slow and tedious. There are two other options. Ask someone else to do it or ask a local DMR operator for a 'code plug' for your radio model, preferably (but not essentially) with your DMR ID on it.

START A CONTACT LIST

If you downloaded a contact list from an Internet site such as RadioID.net or AmateurRadio.digital, you can import it into the CPS software and write it to the radio. See page 24 for the downloading details.

The contact list contains the callsign name and other details of other registered DMR users. You can elect to add contacts yourself making a small contact list containing the callsign and details of people you call often. Or you can download a contact list from online sources.

On the Radioddity CPS, I used **Tool > Import > Digital Contact List** and selected the AnyTone AT-D878UV file that I downloaded from RadioID.net. It imported perfectly.

TIP: You will talk to very few of the 206,000 DMR users in the world, but it makes no difference to the radio whether the contact list has 10 entries or 100,000, so you might as well download as many contacts as will fit in the radio. You certainly want all the contacts from your country or state. It depends a lot on whether you plan to use your DMR for worldwide contacts, nationwide contacts, or local contacts.

Use **Digital > Digital Contact List > 1---20000 etc.** to view the list. You can double click any line on the table to edit the individual entry. Or click on an unused line to add a new contact.

TIP: Leave a few lines free in case you want to add a contact who is not on your list. I did not have room for all of the available DMR IDs, so I didn't choose countries where English is not widely spoken.

When I worked a ham from Indonesia his callsign did not show up, so I added him manually. You can find someone's name, and DMR ID, on Radioid.net by clicking 'Database' and 'DMR User ID Search.' You can also search for repeaters using the 'DMR Rptr ID Search' button. There is more about this on page 151.

ADD SOME TALK GROUPS

Before you add talk groups (TG) you need to decide which DMR network(s) you will be using. Your local repeater will be connected to a DMR network, but other repeaters you access might be connected to a different DMR network. You can set your hotspot to work on any DMR network that you have subscribed to.

My radio will not accept two talk groups with the same TG number, but there are surprisingly few duplicates. When the big networks use the same TG number, it often means that the talk groups serve a similar region. So, you can use the same talk group entry on both networks.

TIP: If you are intending to use multiple networks, for example, a repeater on the DMR-MARC network and a hotspot on the Brandmeister network, you probably should separate them into separate zones. The talk group numbers may mean different things to different networks, and it would be easy to get mixed up. For example, TG 244 on DMR-MARC is Finland, but on the TGIF network, it is the South Devon Radio Club. TG 222 is Italy on DMR-MARC and Brandmeister, Spain on TGIF, and French language on DMR+.

TG 99 is usually allocated to DMR Simplex (radio to radio) operation.

There are three ways to add or edit your talk groups. Option 1 is the best method if you want to have access to a lot of talk groups because you can download or cut-n-paste a list of talk groups from the Internet. Option two is Ok if you just want a few talk groups, and option three can be used when you want to add a single new talk group or make a quick edit. I usually use the method outlined in option 2.

- Option 1: Use a spreadsheet to produce or edit a talk group list
- Option 2: Use the CPS program to create or edit a talk group list
- Option 3: Use the radio keypad to do a quick add or edit

Option 1: Use a spreadsheet to produce or edit a talk group list

You can download a list of talk groups or build your own. Even if you do download a list or cut-n-paste a list into a spreadsheet, you may have to edit it to make it acceptable for your CPS program to import. See editing .csv files below.

A) Download a list of talk groups

Download a list of talk groups for the DMR network(s) you will be using the most.

http://k3nyj.blogspot.com/2019/05/talk_group-lists-dmr-marc-and.html has .csv files for the DMR-MARC and Brandmeister talk groups suitable for the Alinco MD5T and the AnyTone 868 and 878. The latter is OK for the Radioddity GD-AT10G as well. The DMR-MARC file has duplicate talk group numbers which must be removed before the file will upload to the CPS or the radio.

The same blog has TGIF talk groups at http://k3nyj.blogspot.com/2020/06/talk group-list-tgif-network.html

Brandmeister talk groups can be downloaded by clicking 'CSV' at the top of the https://brandmeister.network/?page=talk groups. You will have to do some work to make it suitable for importing into your CPS.

DMR+ talk groups are at https://www.pistar.uk/dmr_dmr+_talk groups.php

Your local amateur radio club or DMR operators may be able to find other sources. Or provide you with a suitable talk group .csv file.

B) Editing .csv files

You can edit a .csv file with Excel, or Open Office spreadsheet, or probably any spreadsheet program. You can even edit it with a word processor like Word or Notepad if you are very careful to make sure that every entry is separated with a comma and every record has a new line. If you use a text editor you must make sure to save the file with a .csv file extension, not the default .txt file extension.

The easiest way to find out what the CPS program will accept is to export the file you want to create. In this case, it is the talk group file. I used Tool > Export > talk groups.

My Radioddity CPS is expecting a file with seven named columns. When you have finished editing the talk group .csv file, save it on your PC and then import it back into the CPS program, using Tool > Import > talk groups. Check out the finished table at Digital > Contact/talk group.

	A	B	C	D	E	F	G
1	No.	Radio ID	Callsign	Name	City	Call Type	Call Alert
2	1	1		Worldwide TG1		Group Cal	None
3	2	3		N America TG3		Group Cal	None
4	3	4		German TG4		Group Cal	None
5	4	5		VK/ZL TG5		Group Cal	None
6	5	8		ZK TG8		Group Cal	None
7	6	9		Local TG9		Group Cal	None
8	7	5300235	ZL3AC	ZL3AC	CHCH ARC	Private Ca	None

Figure 12: Editing a talk group file in Excel

- No. must be consecutive. If you delete, move, or add new lines you will have to renumber the table. Select 1 and 2 at the top, move your mouse to the bottom right of the square that has the 2 in it. Hold down the left mouse button and drag the series down to the last entry.

- If you did it right, dragging the selection will have replaced the numbers with a numerical sequence from 1 to x.

- Radio ID is the talk group (TG) number. It does not have to be consecutive. But it must be the same as the network provider uses, or you will end up at the same party but in the wrong room. The CPS will not upload the file if there are duplicate TG numbers. For private calls, enter the DMR ID of the station you want to include.

- Callsign is only entered for private calls.

- Name is the talk group name. It is less confusing if you leave them the way they were downloaded. But if you have talk groups for two or more networks, I recommend adding a prefix such as BM for Brandmeister, DMR+, or TGIF.

- City is only entered for private calls.

- Call Type is set to Group Call, except for Private Calls and Parrot or Echo, which only you want to hear.

- Call alert sends a beep or vibration to the private call person when you activate the talk group and will return a 'success call' or 'failed call' message to your radio. I have not tried this feature.

C) Group and private calls

DMR network talk groups are marked 'Group Call' because you can call a group of people who have also selected that talk group. The Callsign and City fields are only used if you have added individual contacts to the list. You can set these up as 'Private Call' talk groups, to talk with an individual ham without broadcasting to everyone on a talk group. However, some networks and many repeaters do not support private calling because private calls tie up a timeslot on all involved repeaters making it unavailable for other users. I recommend only using private calls between two stations that are using hotspots rather than repeaters.

D) Trimming the list

The downloaded lists have dozens of talk groups that you don't want. You can delete the talk groups you are never going to use, by editing the .csv file then loading it into the CPS, or by editing the talk group table in the CPS. It is easy to add a new talk group to the list if you change your mind later. Whichever method you choose, you should make sure that the .csv file is the same as the file on the CPS. I have been known to upload an out of date .csv file to the CPS and that is a bad idea!

*TIP: If you are going to trim the list, it is important to do it **before** you allocate talk groups to your Channels. Or you could end up with channels allocated to talk groups that no longer exist.*

When you have finished editing the .csv file, you should renumber the No. column. The numbers do not have to be consecutive. There can be gaps, but not duplicates. I like to index the talk groups on the talk group number then renumber the No column.

It might be the case that you want to use two or more DMR networks. You can download the relevant talk group file, but I recommend caution when adding it to your talk group .csv file. There are some rules.

1. Talk groups cannot share the same TG number. For example, the DMR-MARC list says that TG 5301 is 'New Zealand – North Island,' and Brandmeister says that TG 5301 is 'ZL1 Regional.' These refer to the same regional group on two networks. You can't list them both, but you can use the same talk group entry for channels on both networks. Use different channel names to identify the network.

2. When you are naming the talk groups for an additional DMR network, it is not a bad idea to add a prefix or suffix, to avoid confusion. For example, I add DMR+ to all my DMR+ talk groups and TGIF to my TGIF talk groups. You don't have to do this, but each time you create a channel for a repeater you have to pick a talk group and it is easier if the talk groups are easy to identify.

E) Save and upload

Save the .csv file as 'AnyTone TG.csv' or 'Radioddity TG.csv' and import the file into the CPS software using **Tool > Import > Talk Groups.**'

Save the code plug using the **Save** icon or **File Save.**

Connect the radio and upload the file to the radio. See 'Upload and download data from and to an AnyTone/Radioddity radio' on page 31.

Option 2: Use the CPS program to create or edit a talk group List

You can add, edit, or delete talk groups directly in the CPS program. This is the method I usually use. It is quicker if you are only adding a few talk groups.

A) Add or edit a talk group in the CPS

At a minimum, you need to create talk groups for

- 'Local' on TG 9. TS1 is used when you want to talk with someone over the local repeater, but don't want the call to be broadcast to other linked repeaters. TG 9 TS2 is used to talk on a repeater that is connected to a reflector.

- 'Regional' on TG 8 TS1. TG 8 is often linked to a regional talk group that is only available within your immediate DMR network. On my hotspot, I use TG 8 TS2 to talk on the DMR+ reflectors. Pi-Star converts it to TG 9 TS2.

- 'Parrot' or 'Echo' on TG 9990

- 'Disconnect' on TG 4000 if you plan on adding any dynamic talk groups. TG 400 on Phoenix UK

- and 'Simplex' on TG 99

You also need the static talk groups that are available on the repeater. In fact, for all the repeaters that you plan to access. All repeaters in your region that are on the same DMR network will probably have the same or similar static talk groups. See the 'Finding repeater information' chapter on page 133 for details on how to find the static talk groups for your local repeaters.

Finally, you can add the dynamic talk groups that you are interested in. They must match the DMR network of the repeater or hotspot, that you are planning to access.

Let's start by adding 'Local', or 'LCL' on TG 9. This talk group is necessary for repeaters and for accessing reflectors on DMR+ hotspots.

On the Radioddity/AnyTone CPS, the talk group list is at, **Digital > Contact/talk group**. The columns are No., TG/DMR ID, Call Alert, Name, and Call Type. There may be talk groups listed already, or there may just be a couple of dummy talk groups that can be replaced.

Check that TG 9 does not already exist then double click anywhere on a blank line. A popup box will appear. Enter **Local TG 9** in the Name field, and **9** in the TG/DMR field. The Call Type dropdown box should be set to **Group Call**, and the Call Alert will be greyed out. Press **OK** to save. Yipee, that's your first talk group! Repeat this for other talk groups on your network such as TG 91, TG 505, TG 235 etc.

The process for Parrot or Echo is similar. Double click the next blank line. In the Name field, enter 'Parrot' for the Brandmeister network or 'Echo' if the repeater is linked to the DMR+ network. Enter **9990** in the TG/DMR field. This time the Call Type dropdown box should be set to **Private Call**. Press OK to save. Yipee, that's your 2nd talk group!

We will do one more and then you're on your own. This time we will enter a Private Call contact. Double click a blank line and enter the **name and callsign** for the contact in the Name field. It might be another ham, or a club call etc. Enter the user's **DMR ID** in the TG/DMR field. If they have been active on DMR for a while, their name and ID will probably be in your contact list. Or you can look it up by entering their callsign on RadioID.net. The Call Type dropdown box should be set to **Private Call**. The Call Alert can be set to None, Ring, or Online Alert. Call alert causes a ring or vibration on the radio belonging to the person you want to call. It will return a 'success call' or 'failed call' message to your radio. I have not tried this feature.

Option 3: Use the radio keypad to do a quick add or edit

If you don't have a computer and programming cable handy, or you just want to add one more talk group or make a quick edit. You can "do things the hard way."

TIP: You enter text using the ABC layout on the keyboard. One press highlights A, a second press highlights B etc. the same as using numeric the keyboard on an old phone. The red - - key is backspace delete. The * key enters a space. The # key changes the entry mode from uppercase, to numbers, to lowercase. This is indicated in the top right of the entry screen.

- Pressing the red Back (- -) key repeatedly backs you up through the menu structure until 'Back' changes to 'Exit' and you end up back on the main screen.

To edit a talk group using the radio keypad

- Click the **Menu** (-) button

- Use the round **up-down** button to highlight **Talk Group** click **Menu** (-)

- Click the **Menu** (-) button again to select **TG List**. You will see all of the talk groups. Use the round **up-down** button to highlight the talk group that you want to edit, and press **Select**. You get the choice to **View**, **Edit**, or **Select** the talk group or contact.

- If you select Edit, you can edit the

    1. ID (talk group number or DMR ID for private calls)

    2. Name (talk group name)

    3. Address, callsign, state, country, and remarks (private calls only)

To add a talk group using the radio keypad

- Click the **Menu** (-) button

- Use the round **up-down** button to highlight **Talk Group** click **Menu** (-)

- Use the round **up-down** button to highlight **New Contact** click **Menu** (-)

- Highlight each option, in turn, using the **up-down** button then click **Menu** (-) to select it. Enter the required information.

    1. ID (talk group number or DMR ID for private calls)

    2. Name (talk group name)

    3. Address, callsign, state, country, and remarks (private calls only)

- Note, the radio will not let you save an entry if the talk group ID already exists. If the talk group already exists, you can allocate it to a channel in the next section.

PROGRAM SOME CHANNELS

Programming a channel is more complicated than programming a channel on an FM radio. For a start, you need a channel for every talk group that you want to use on every repeater that you want to use. This means that you have to create many channels for each repeater frequency pair. I don't know why the CPS software makes this such an ordeal, I would have thought that you could create one RF channel and then allocate multiple talk groups to it... but unfortunately, it does not work that way. You also need to allocate the correct time slot and colour code to each channel. The colour code will be the same for all channels on a particular repeater, but the time slot has to be adjusted to suit the talk group.

Each channel must have the correct frequencies, colour code, and timeslot. As usual, you can program channels directly in the CPS software, or by creating a .csv file and uploading it into the CPS radio or using the radio keypad buttons on the radio. But the 'channel.csv' file has 51 columns, so I found that entering each channel directly into the CPS is the best option.

If someone can send you a suitable channel file or you can download one, then that makes spreadsheet entry a viable option. It is also a good way of rearranging your existing channel list into a more logical order. Entering channels from the radio keypad is pretty slow, but OK for adding one or two new channels. There are some notes about finding the right channel time slot and colour code, in 'Check that you have the right settings' on page 63. Following that there is a brief note describing what time slots and colour codes are.

- Option 1: Use the CPS to make a channel list (preferred method)
- Option 2: Use a spreadsheet to produce a channel list
- Option 3: Use the radio keypad to do a quick add or edit

Option 1: Use the CPS to make a channel list

I believe that entering channels directly into the CPS is much less confusing than using the spreadsheet entry method. You have to create a channel for every talk group that you want to use on the specific repeater or hotspot. The talk groups you want to select must already be in your talk group list.

A) Before you start

You will need to know the repeater or hotspot's transmit frequency, receive frequency or offset, colour code, DMR network, and static talk groups (if any).

In the CPS, click **Common Setting > Channel** to bring up the channel list.

B) Program an FM channel

We will start with an FM channel for an analog repeater. You only need one channel for each FM repeater. Skip this step if the radio will only be used for DMR.

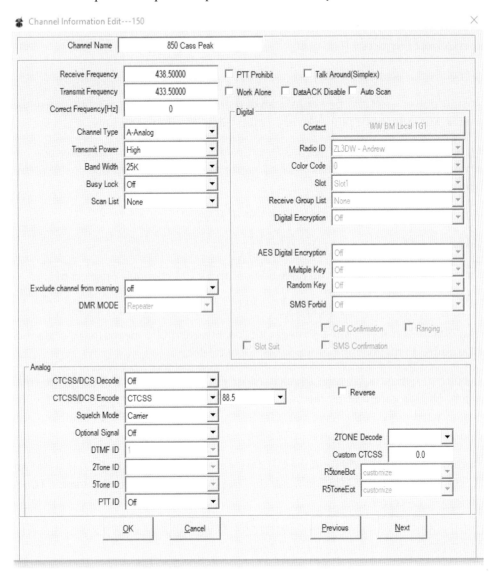

Figure 13: FM channel setup screen

Channel name – enter a channel name that identifies the repeater. In New Zealand that consists of the repeater name 'Cass Peak' and a code representing the frequency. The repeater transmits on **438.50** so the code is 850.

Receive frequency – set this to the repeater output frequency.

Transmit frequency – set this to the repeater input frequency.

Leave the five checkboxes on the right unchecked.

TIP: The 'Talk Around (Simplex)' checkbox sets the transmitter to the stated receiver frequency. It can be used if you want to temporarily transmit on the output frequency of a repeater.

Leave Correct Frequency [Hz] at **0**

Channel Type – set to **A-Analog**

Transmit power – set the power to a level that ensures you can reliably access the repeater. My radio has 4 power levels, Turbo, **High**, Medium, and Low.

TIP: To extend the battery life, use the lowest power possible. Higher transmit power can dramatically increase battery drain. I would normally set the Transmit Power to High (not Turbo) for repeaters.

Bandwidth is always **25 kHz** for analog FM channels

Busy Lock – **off**

Scan list – select a scan list if you have one set up for FM channels or select **off**.

APRS report type is usually set to **off** for FM

Exclude channel from roaming – this is a bit "backwards." **Off** means that the channel can be added to the roaming group i.e. is not excluded.

The bottom section is for tone squelch. Most FM repeaters use CTCSS, or no tone squelch. DCS is rare and as far as I know, DTMF is not used anywhere. If your repeater needs a CTCSS tone, set CTCSS/DCS Encode to **CTCSS** and use the dropdown list to select the correct **CTCSS tone** for the repeater.

Generally, the repeater will send the CTCSS tone on the output as well, so you can set CTCSS/DCS Decode to **CTCSS** and use the dropdown list to select the **same CTCSS tone** for your receiver.

Squelch mode - If you have selected a CTCSS decode tone, change the dropdown to **CTCSS/DCS**, otherwise leave it set to **Carrier**.

Leave 'Optional Signal' **off**, DTMF ID, 2Tone ID, 5tone ID **not set**, and PTT ID **off**.

Leave, 'Reverse' **unchecked** and ignore the four boxes below that.

C) Program an FM simplex channel

An FM simplex channel is the same as an FM repeater channel, except the receive and transmit frequency will be the same, and you would not normally use CTCSS tones. Set CTCSS/DCS Encode and CTCSS/DCS Decode to **Off**.

D) Program a DMR channel

You should program a channel for every talk group that you want to use, on each repeater or hotspot.

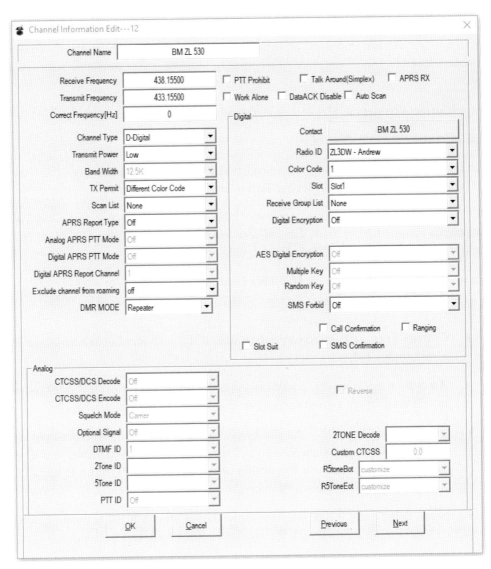

Figure 14: DMR channel setup screen

Channel name – enter a channel name that identifies the talk group and possibly the repeater. How you do this depends on how you plan to arrange your zones. If your zone includes several repeaters, you should identify the repeater in the channel name. But if the zone only includes one repeater or hotspot, you don't need to include the repeater name because it will already be in the zone name. In this example, the channel is on my hotspot for the ZL talk group on Brandmeister TG 530.

Receive frequency – set this to the repeater or hotspot output frequency.

Transmit frequency – set this to the repeater or hotspot input frequency. Both frequencies will be the same if you have a simplex hotspot.

Leave the five checkboxes on the right **unchecked**.

TIP: The 'Talk Around (Simplex)' checkbox sets the transmitter to the stated receiver frequency. It can be used if you want to temporarily transmit on the output frequency of a repeater.

Leave 'Correct Frequency [Hz] at **0** unless you have a reason not to.

Channel Type – set to **D-Digital**

Transmit power – set the power to a level that ensures you can reliably access the repeater. My radio has 4 power levels, Turbo, High, Medium, and Low.

TIP: To extend the battery life use the lowest power possible. Higher transmit power can dramatically increase battery drain. I would normally set the Transmit Power to 'High' (not Turbo) for repeaters and to 'Low' for a hotspot.

Bandwidth is locked. It is always **12.5 kHz** for digital channels

TX permit

- **Channel Free** - the radio can only transmit if the channel is available. Good for a simplex hotspot channel.

- **Always** – there is no constraint on when the radio can transmit. Could cause interference but is a good option for your TG 4000 disconnect channel.

- **Different color code** – the radio can transmit even if the transmit colour code does not match the received colour code. But the frequencies and time slot must be correct. I use this option because it allows the radio to transmit when the hotspot is transmitting on the alternate time slot.

- **Same color code** – the radio can only transmit even if the transmit colour code and time slot do match the received colour code and time slot.

Scan list – select a scan list if you have one set up for DMR channels or select off.

APRS report type is usually set to **off** for DMR channels. See the chapter on using APRS, page 127.

Exclude channel from roaming. This is a bit "backwards." **Off** means that the channel can be added to the roaming group i.e. is not excluded.

DMR mode – this must be set to **Repeater** for repeaters or duplex hotspots and **Simplex** for simplex hotspots or digital simplex channels.

All the above information will remain the same for all of the channels allocated to the repeater or hotspot. The 'Digital' section allocates the talk group to the channel.

TIP: Some people only program two channels. One for a talk group on TS1 and the other for a talk group on TS1. If you leave the Digital Monitor turned on (one slot), you will hear all the static talk groups on the time slot, plus any dynamic or auto-static talk group. If you want to call a station, take a note of the displayed TG and select it using the radio buttons. I prefer to have individual channels for my talk groups, but it is a valid approach especially if you normally only transmit on one or two talk groups anyway. Change the digital monitor to 'double slot' if you want to hear both time slots.

Contact – this is where you set the talk group, (or digital ID for private calls). Click the button and you will see all the talk groups in your talk group list. Double click the one that you want to associate with the channel. If you have talk groups for two or more DMR networks, you will see why it is helpful to prefix your DMR network talk groups with BM, DMR+, or TGIF. If two networks use the same TG number for the same talk group, you can select it even if it has the wrong name.

Radio ID – will **default** to your DMR ID name and would only need to be changed if you are using multiple DMR IDs on the radio.

Color Code – must be set correctly for the repeater or hotspot. It will be the same for all channels on the repeater. It is one of the repeater facts you need to gather. See 'Finding repeater information' on page 133. If in doubt set it to **1**.

Slot - the repeater provider should also provide information regarding what time slot to select for each talk group. The time slot for the static talk groups can be gathered from the IPSC2 or Brandmeister dashboard, or the repeater dashboard or website. See 'Finding repeater information' on page 133. You can also use the 'digital monitor' to find out what time slots are associated with each talk group. See 'Check that you have the right settings,' on page 63.

TIP: If you are using a simplex hotspot all channels will be on TS2. If you are using a duplex hotspot, you will probably have allocated TS1 to local talk groups and TS2 to overseas talk groups. Select the 'Slot' accordingly.

Receive group list - sets the receive group list that will be active when you have the channel selected. Usually **none**. See 'receive group list' on page 68.

TIP: it is possible to hear talk groups that are in the receive group call list that you have chosen, even if you have not made a channel for that talk group. This can leave you in a confusing situation where you can hear a caller but not be able to transmit to them. For example, let's say that TG 91 is a static talk group on the repeater. You have added TG 91 to a receive group call list called 'World." You create channel number 1 for TG 123, and you select 'World' in the receive group call list dropdown. When you have the radio set to channel 1, you will be able to hear TG 91 even though you don't have that talk group programmed for the repeater. You will also hear callers on TG 123 and will be able to transmit back to them.

Digital Encryption – not allowed in most countries. Set it to **Off**.

AES Digital Encryption, Multiple Key, and Random Key will be **greyed out** because you don't have encryption set up.

SMS forbid – stops the radio from sending SMS messages on the channel. Including APRS. Leave it turned **off**.

Leave the four checkboxes **unchecked**.

Click **OK** to save and exit.

E)   Program a subsequent DMR channel

Well done! You have programmed your first DMR channel. The good news is that the second one will be easier.

In the channel list, single click on the channel that you just finished. Right-click and select 'Copy' or use CTRL-C to copy the line onto the clipboard. Click on the next blank channel. Right-click and select 'Paste' or use CTRL-V to paste the line into the channel data. That makes a copy of the first channel, so you don't have to enter all the information again.

Double click on the new line. All the left side frequency information will be fine. All you need to do is write another **Channel name** and select another **talk group**. Check that the time **Slot** is correct, the **Scan** setting, and the **receive group list**. Then press **OK** to exit and carry on with the next channel.

F)   Program a digital simplex channel

A digital simplex channel is the same as a digital repeater channel, except the receive and transmit frequency will be the same. Set DMR Mode to **Simplex** and select the **TG 99** simplex talk group. Set the **time slot** to **TS1**. The **colour code** must be the same on both radios, but the time slot does not matter. The radio will ignore the slot setting on Simplex (but not on Double Slot).

G)   Save and upload

Save the code plug using the **Save** icon or **File Save**.

Connect the radio and upload the file to the radio. See 'Upload and download data from and to the radio (AnyTone/Radioddity)' on page 31.

Option 2: Use a spreadsheet to produce a channel list

You probably will not be able to find a channel list online, but your local DMR group or radio club might have one, or another DMR user. The easiest way to find out what the CPS program will accept is to use the CPS to enter one channel entry or all the channels for one repeater and then export the 'channel' file. Then enter the remaining channels using a spreadsheet. I used Tool > Export > Channel.

My Radioddity CPS is expecting a file with fifty-one named columns. Most of them can simply be copied down the list.

When you have finished editing the talk group .csv file, save it on your PC and then import it back into the CPS program, I used Tool > Import > Channel. Check out your finished table at Common Setting > Channel.

No.	Channel Name	Receive Fi	Transmit F	Channel T	Transmit F	Band Widt	CTCSS/DC	CTCSS/DC	Contact	Contact Call	Contact T(
4	BM WW 91	438.125	433.125	D-Digital	Low	12.5K	Off	Off	BM WW	Group Call	91
5	BM EU 92	438.125	433.125	D-Digital	Low	12.5K	Off	Off	BM EU	Group Call	92
6	BM WW 93	438.125	433.125	D-Digital	Low	12.5K	Off	Off	BM North America	Group Call	93
10	Canada wide 302	438.125	433.125	D-Digital	Low	12.5K	Off	Off	Canada Wide	Group Call	302
34	DMR+ UK 235	438.125	433.125	D-Digital	Low	12.5K	Off	Off	DMR+ UK	Group Call	80235
35	DMR+ VK 505	438.125	433.125	D-Digital	Low	12.5K	Off	Off	DMR+ Australia	Group Call	80505
36	DMR+ Oceania 5	438.125	433.125	D-Digital	Low	12.5K	Off	Off	DMR+ Oceania	Group Call	80005
37	DMR+ ZL 530	438.125	433.125	D-Digital	Low	12.5K	Off	Off	DMR+ ZL	Group Call	80530
45	TGIF 110	438.125	433.125	D-Digital	Low	12.5K	Off	Off	TGIF 110	Group Call	5000110
47	TGIF WW 31226	438.125	433.125	D-Digital	Low	12.5K	Off	Off	TGIF WW 31226	Group Call	5031226
48	TGIF Skynet 3703	438.125	433.125	D-Digital	Low	12.5K	Off	Off	TGIF Skynet 3703	Group Call	5037030

Figure 15: The most important columns in the Channel .csv file

You will see the same pattern of channels for each repeater and hotspot. Each repeater needs a channel for each of the static talk groups, the dynamic talk groups you might use, local, disconnect, and Parrot or Echo.

You can cut-n-paste whole rows, or select and drag many of the entries down the list as many of the entries are the same on each line. Some settings in the spreadsheet cannot be changed in the CPS and are probably not relevant to the radio model.

Channel Name – is usually a combination of the talk group name and the channel name. Whatever is going to be easy to remember. See the image above. If the channel will only be used in one zone, you could omit the 'BM,' 'DMR+' prefix and just put the channel in a BM or DMR+ zone. 'Canada Wide' has no prefix because 302 is Canada on all DMR networks.

Receive frequency is the DMR repeater or hotspot output frequency

Transmit frequency is the DMR repeater or hotspot input frequency

Channel Type - these are D-Digital channels. Another repeater could be set up for A-Analog channels.

Transmit Power – to save battery power use as little as possible that ensures reliable communications. I recommend **Low** for a hotspot at your home, **High** for repeaters, and **Turbo** for distant repeaters.

Bandwidth – always **12.5 kHz** for DMR and **25 kHz** for FM

CTCSS/DCS Decode – is for FM repeaters only enter the CTCSS Tone

CTCSS/DCS Encode – is for FM repeaters only enter the CTCSS Tone

Contact is the talk group name. It must be exact, or it will not be accepted. So, if possible, copy-n-paste it from a previous entry.

Contact Call – is **Group Call** for all talk groups except Parrot/Echo. Use **Private Call** for private calls where you are calling a DMR ID rather than a talk group. In the CSP program, this is set automatically when you select the talk group

Radio ID – is the name that you used when you entered your DMR ID

Busy Lock/TX Permit – there are several options.

- The CPS says that **Channel Free** is the usual setting. The radio can only transmit if the channel is available

- **Always** – there is no constraint on when the radio can transmit. Don't use.

- **Different color code** – the radio can transmit even if the transmit colour code does not match the received colour code. But the frequencies and time slot must be correct. **I always use this option**.

- **Same color code** – the radio can only transmit even if the transmit colour code and time slot do match the received colour code and time slot.

Squelch Mode – is for FM channels only. Can be set to Carrier, or CTCSS if a CTCSS tone has been set. For DMR channels set it to Carrier and copy it all the way down the list.

Optional Signal – is for FM channels only

DTMF ID   – is DTMF tone selection for FM channels only

2Tone ID   – is DTMF tone selection for FM channels only

5Tone ID   – is DTMF tone selection for FM channels only

PTT ID   – is DTMF tone ID on PTT for FM channels only

Color Code – must be set correctly for the repeater or hotspot. It will be the same for all channels on the repeater. It is one of the repeater facts you need to gather. See 'Finding repeater information' on page 133.

Slot - the repeater provider should also provide information regarding what time slot to select for each talk group. The time slot for the static talk groups can be gathered from the IPSC2 or Brandmeister dashboard, or the repeater dashboard or website. See 'Finding repeater information' on page 133. You can also use the 'digital monitor' to find out what time slots are associated with each talk group. See 'Check that you have the right settings,' on page 63.

Scan List – adding the Scan List to the channels seems to not affect the channels listed in the Scan List section or the scan list in the radio.

Receive group list - sets the receive group list that will be active when you have the channel selected. Usually **none**. See 'receive group list' on page 68.

PTT Prohibit – stops the channel transmitting. Usually, **off**.

Reverse – reverses the TX and RX channels for FM channels only

Simplex TDMA - unknown use. It seems to be 'On' for analog channels and 'Off' for digital channels.

Slot Suit – makes the radio ignore the time slot setting when receiving in 'Repeater' mode. Set to **off**.

AES Digital Encryption - always set to '**Normal Encryption**'

Digital Encryption – not allowed. Set to **off**.

Call Confirmation – usually **off**. Blocks Private Calls unless your radio has transmitted on the contact ID (private call talk group).

Talk Around (Simplex) makes the RX frequency the same as the TX frequency. Usually, **off**.

Work Alone – enables the 'work alone' safety function. It is usually only used in commercial radio systems.

Custom CTCSS - for FM using a non-standard CTCSS tone. I can't see any reason to use this.

2TONE Decode - for FM. Squelch using a received DTMF pair. I don't think this is supported on my radio, and you wouldn't use DTMF squelch anyway.

Ranging – This seems to be a polling function between two radios operating in simplex mode. Because your radio will poll the other radio this function should not be used on repeaters or talk channels. Turn it on to display the range between two GPS equipped radios working simplex radio-to-radio. GPS must be turned on in your radio for this to work. I have never made this work.

Through Mode – unknown feature set to **Off**. Seems to be related to the Simplex / Repeater setting

APRS RX - Turn **'On'** to display APRS beacons from the APRS talk groups. GPS and APRS must be turned on in your radio for this to work.

Analog APRS PTT Mode – not settable in the CPS

Digital APRS PTT Mode – not settable in the CPS

APRS Report Type – not settable in the CPS

Digital APRS Report Channel – not settable in the CPS

Correct Frequency [Hz] - a frequency adjustment to get the radio exactly on frequency

SMS Confirmation. I don't know what this does so I leave it **unchecked**

Exclude channel from roaming – excludes the channel from the roaming list

DMR MODE - must be set to **1** (Repeater) for repeaters or duplex hotspots. Set it to **0** (Simplex) for simplex hotspots or digital simplex channels. 2 is for 'double slot' which we don't use. 'Repeater' is used for duplex hotspots and repeaters that have different input and output frequencies. 'Simplex' is used when the same RX and TX frequencies are used, to communicate directly with another radio. Simplex can be used with two frequencies so long as the other radio reverses the frequency pair. 'Double slot' is used when you receive and transmit on the same frequency. The time slot must be the same.

DataACK Disable - **0**, copy down the whole list. Makes the radio not respond to information requests from the repeater. Probably for the call alert.

R5toneBot - **0**, copy down. FM using DTMF tones.

R5ToneEot - **0**, copy down. FM using DTMF tones

Auto Scan - **0**, copy down.

Ana Aprs - **0**, copy down

Mute – **blank**, copy down

Save and upload

Save the channel list as a .csv file. Then use **Tool > Import > Channel** to import it back into the CPS. Then save the code plug using the **Save** icon or **File Save**. You can check the channel list looks ok and if necessary, make any minor changes, using **Common Setting > Channel**. If you are happy, connect the radio and upload it to the radio.

Option 3: Use the radio keypad to do a quick add or edit

If you don't have a computer and programming cable, or you just want to add one more channel or make a quick edit. You can "do things the hard way."

TIP: You enter text using the ABC layout on the keyboard. One press highlights A, a second press highlights B etc. the same as using numeric the keyboard on an old phone. The red - - key is backspace delete. The * key is a space. The # key changes the entry mode from uppercase, to numbers, to lowercase. This is indicated in the top right of the entry screen.

- Pressing the red Back (- -) key repeatedly backs you up through the menu structure until 'Back' changes to 'Exit' and you end up back on the main screen.

To add a channel using the radio keypad

This creates a channel that is the same as the channel that the radio was sitting on when you started the process. So, select a channel that is close to what you are trying to add. Preferably one on the same repeater or hotspot. After the new channel has been added, you will have to select it, and then edit it, to change parameters. Selecting a different talk group at the very least.

- Click the **Menu** (-) button

- Use the round **up-down** button to navigate down to **Settings** and click **Menu** (-) to 'Select.'

- Click down to option 2 **Chan Set** and click **Menu** (-) to 'Select.'

- On option 1 **New Chan**, click **Menu** (-) to 'Select.'

- The radio will select the next available channel. You can change it if you want to. Then click **Menu** (-) to Confirm

- Enter the channel name and click **Menu** (-) to Confirm

- Select a Zone from the displayed list click **Menu** (-) **Menu** (-) to Confirm and save.

- This creates a channel the same as the channel that the radio was sitting on when you started the process. You will need to go on to the editing process below to change the talk group. The other parameters should be OK.

To edit a channel using the radio keypad

- Set the radio to the zone and channel that you want to edit

- Click the **Menu** (-) button

- Use the round **up-down** button to navigate down to **Settings** and click **Menu** (-) to 'Select.'

- Click down to option 2 **Chan Set** and click **Menu** (-) to 'Select.'

- There are 22 settings that you can change

    3) Channel Type. A-Analog or **D-Digital**

    4) TX power – turbo, high, medium, **low**

    5) Offset – **zero** unless you have a reason to change it

    6) Bandwidth – the only option for digital is **'narrow,'** (12.5 kHz). Use wide (25 kHz) for FM

    7) RX frequency. You can change the frequency and there are two options Confirm and Delete. To exit the screen press P2.

    8) TX frequency. You can change the frequency and there are two options Confirm and Delete. To exit the screen press P2.

    9) Talk around – **off** or on. Changes the radio from repeater to simplex mode. I suggest you don't use it.

    10) Name. You can change the name using the alpha keys and there are two options Confirm and Delete. To exit the screen press P2.

    11) TX Allow; always, channel free, **different CC**, same CC.

    12) TX prohibit on or **off**

    13) Radio ID – usually there is only one, **your DMR ID**

    14) Color Code – select 0 – 16. Default = **1**

    15) Time Slot – select **TS1 or TS2**

    16) Digital Encryption should be **off**

    17) Encrypt Type – does not matter

    18) RX Group List – select or add a receive group list

    19) Work alone – not usually used

    20) CH ranging on or **off**

    21) APRS receive on or **off** (GPS and APRS needs to be turned on)

22) SMS forbid on or **off**. Usually off which means SME messaging can be used.

23) Data ack forbid – usually **off**

24) DMR Mode **Simplex** for Simplex hotspots and digital simplex channels. **Repeater** for repeaters and duplex hotspots.

Check that you have the right settings

If you have the DMR channel programmed and you can see transmissions on the RSSI (receive signal strength indicator), but you cannot hear anything. You may be listening to a different talk group, or you might have the wrong time slot or colour code selected. Use the 'digital monitor' (sometimes called promiscuous mode) to find out the talk group, colour code and time slot for the repeater. You can turn on the 'digital monitor in the CPS. Select **Common Setting > Optional Setting > Digital Func > Digital Monitor** and select **Double Slot**. On the same page, set **Digital Monitor CC** to **Any** and **Digital Monitor ID** to **Any**. When a call is made the display will show the TG that is in use, the TS (time slot) and the CC (colour code). Once those settings have been made you can click the PF2 button once to turn on single time slot monitoring and a second time to turn on dual time slot monitoring, the third click turns the monitor off.

Colour Code (CC)

The colour code is used to identify the output of a specific repeater, much like the CTCSS tone on an FM repeater. In the unlikely event that two repeaters near you are using the same frequencies. Using different colour codes will ensure that you only hear the wanted repeater. If in doubt, try CC01. I have been told that all VK and ZL repeaters use CC01. If you have set the colour code incorrectly you will not be able to hear the repeater. [Unless you have the digital monitor – dual slot enabled].

Time Slot (TS)

A repeater or duplex hotspot transmits two time slots. Your channel must have the right one or you will not be able to hear or talk on the wanted talk group. A simplex hotspot only transmits on TS2.

ESTABLISH SOME ZONES

OK, you have some talk groups (contacts) and you have set up some channels on the local repeater or your hotspot. There is one more thing that you have to do, and that is to establish at least one zone. Zones are a bit like the memory groups on an FM radio. They can be named, and you can use them in various ways. If you travel to another city for work and will be using different repeaters you could have a Zone for 'home' and another for 'work.' Or you could have a zone for your local repeater and a different one for your hotspot and perhaps a third for FM repeaters.

After you have entered the channels, you can place them into zones. When the zone is selected you will only be able to choose channels that are in it. If you travel to another city, you will select a different zone and be able to reach the appropriate repeaters. I have a zone for my hotspot, one for a local repeater that is on the Brandmeister network, and another for a local repeater that is on the DMR-MARC/DMR+ network.

The Radioddity GD-AT10G can store up to 250 Zones. You can import Zones from a .csv file using Tool > Import > Zone. But I think that it is much easier to add them directly in the CPS under Common Setting > Zone. Of course, you can also, "do it the hard way," directly from the radio keyboard.

The TYT MD-UV380 and MD-UV390 can have up to 250 Zones with up to 64 channels in each. The older TYT MD-380 and MD-390 models can only have 16 channels per zone. In my opinion, that's plenty.

Use the CPS to create some zones

I believe that entering zones directly into the CPS is much less confusing than using the spreadsheet entry method. You must have at least one zone. I currently have six.

A) Add a new zone (or edit an existing zone)

In the CPS under Common Setting > Zone. Double click a blank line, or an existing line if you want to edit a zone.

Zone name – enter a name for your zone. The one shown on the next page is for my hotspot on the Brandmeister DMR network.

A Channel – enter the channel you will use the most (not critical)

B Channel – enter a channel you would like to have on the second display when you have two channels showing on the radio screen, (not critical)

Order by – at the bottom of the left list. You can arrange the talk groups in order of the ID (talk group or contact number), or the talk group or contact name.

Navigate down the left list to the channels you want in the zone. They will often be bunched together because you will have created them in a group. This is where prefixing can help. Or just pop into the channel list and make a note of the channel numbers you want.

Click >> to move a channel into a zone. Click >> again to move the next channel into the zone. Click >> again to move the next channel into the zone.

Select in the right side list and click << if you want to take a channel out of the zone.

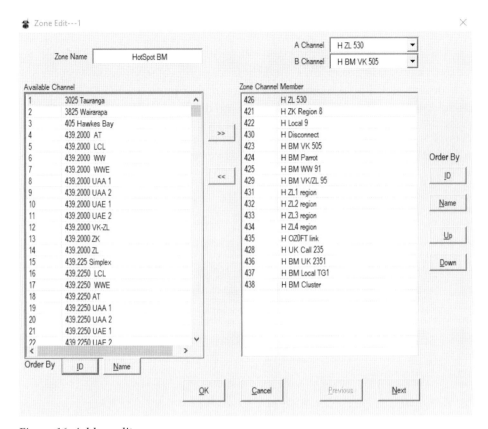

Figure 16: Add or edit a zone

Right side **ID** button - arrange the selected talk groups, in the right side list, in order of the talk group number.

Right side **Name** button - arrange the selected talk groups, in the right side list, in order of the talk group name.

Right side **Up** button - move the selected talk groups up the list to arrange the channels in a logical order on your radio. This sets the order that the channel switch on the top of the radio selects the channels.

Right side **Down** button - move the selected talk groups down the list to arrange the channels in a logical order on your radio.

B) Save and upload

Save the code plug using the **Save** icon or **File Save**.

Connect the radio and upload the file to the radio. See 'Upload and download data from and to the radio (AnyTone/Radioddity)' on page 31.

ADD CHANNELS TO YOUR SCAN LIST

My radio supports up to 250 scan lists. I have set up two, which I never use. You can add any channels to a Scan List, but scanning is better suited to FM.

If you want to listen to all of the static talk groups and the currently selected auto-static or dynamic talk group, on a single hotspot or repeater, you can select the single slot or double slot digital monitor (also known as 'promiscuous' mode). However, you could make a scan group that scans several different DMR repeaters.

*TIP: If the scan stops on an active signal, you can press the green **Menu** button to stop the scan from resuming and stay on that channel.*

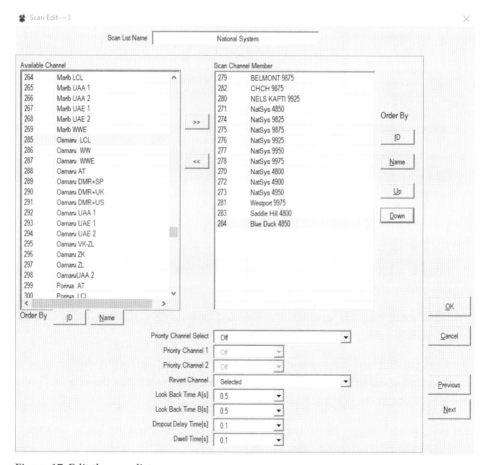

Figure 17: Edit the scan list

Scan List Name – enter a name for your scan list. The one shown above is for the New Zealand network of linked FM repeaters.

Navigate down the left list to the channels you want in the scan list. They will often be bunched together because you will have created them in a group. This is where prefixing can help. Or just pop into the CPS channel list and make a note of the channel numbers you want.

Click >> to move a channel into a scan list. Click >> again to move the next channel into the scan list. Click >> again to move the next channel into the scan list.

Select in the right side list and click << if you want to take a channel out of the list.

Right side **ID** button - arrange the selected talk groups, in the right side list, in order of the talk group number.

Right side **Name** button - arrange the selected talk groups, in the right side list, in order of the talk group name.

Right side **Up** button - move the selected talk groups up the list to arrange the channels in a logical order on your radio. This sets the order that the channel switch on the top of the radio selects the channels.

Right side **Down** button - move the selected talk groups down the list to arrange the channels in a logical order on your radio.

Priority channels

I have not bothered setting priority channels because the radio is dedicated to DMR and I don't use the scan function.

These are functions that you would change from the radio keyboard if you wanted to alter the way the radio scans.

Priority Channel Select – off, or the priority channel 1 selected below, or the priority channel 2 selected below, or both priority channels.

Priority Channel 1 – a dropdown to select a channel as priority channel 1.

Priority Channel 2 – a dropdown to select a channel as priority channel 2.

Revert channel selects what happens if you press the PTT during scanning.

- Selected – transmits on the channel the radio was on before you started the scan
- Selected + Talk Back – if the scan has not stopped it transmits on the channel the radio was on before you started the scan. If the scan has stopped on an active signal, it transmits on the active channel
- Last called – if the scan has not stopped it transmits on the channel that heard a call

- Last Used – if the scan has not stopped it transmits on the last channel that you used.

- Look back time A – jumps back to scan the priority channel(s) every 0.5 to 5 seconds during the scan.

- Look back time B – FM only. Jumps back to scan the priority channel(s) every 0.5 to 5 seconds during the scan. Even if the CTCSS tone is wrong.

- Dropout delay time - FM only. Sets the time after you finish transmitting before the radio starts to scan again. 0.1 to 5 seconds.

- Dwell time - FM only. Sets the time after you finish transmitting before the radio starts to scan again. 0.1 to 5 seconds. Yes, this seems to be identical to the line above!

Save and upload

Save the code plug using the **Save** icon or **File Save**.

Connect the radio and upload the file to the radio. See 'Upload and download data from and to the radio (AnyTone/Radioddity)' on page 31.

PROGRAM SOME RECEIVE GROUP LISTS

You do not have to use receive groups on the AnyTone, BTECH, Radioddity radios. If you do decide to create receive groups they should be planned with care. They work a little like the digital monitor (promiscuous) mode, but they are not as easy to turn off. If you associate a receive group list with a channel, you will hear the calls on the talk groups in the receive group as well as the talk group allocated to the channel. It is great to check if there is activity on the repeater. But it can be confusing. For example, you hear someone make a call, but you don't know what talk group to respond on. You can include dynamic talk groups in the list, but you will only hear them if you or another operator transmits and activates the talk group for 15 minutes.

TIP: Some old radios require the use of a receive group list on every channel to hear anything. On most other radios, if you do not select a receive group list when you create a channel, you will only hear the talk group associated with that channel.

My preferred method

It is a personal preference, but I believe that it is a bad idea to allocate a receive group list on every channel. I think that it is better to create two additional "monitor channels" and link them to two receive group lists. One for each time slot. You will be able to switch to the 'TS1 Monitor' channel and hear all the static talk groups on TS1. Or switch to the 'TS2 Monitor' channel and hear all the static talk groups on TS2. When a call comes in which you want to respond to, you switch channels to the correct talk group, then make your call.

Note if you are using a simplex hotspot, everything is on TS2, so you only need one monitor channel.

TIP: The monitor channel still has to be assigned to a talk group, with a time slot, and a colour code. Use the busiest channel in the receive group list because it will be the one you hear most often. For example, if you have BM TG 91 set as a static talk group, set the 'Contact' to that and you will not have to QSY to another channel to respond to calls in TG 91.

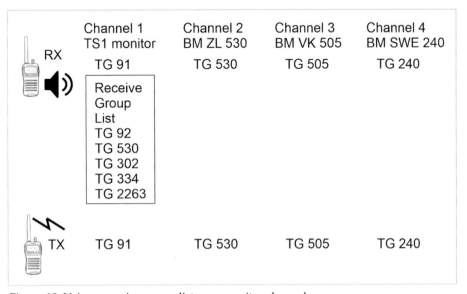

Figure 18: Using a receive group list on a monitor channel

Rules governing receive groups

1. The radio will only hear talk groups that are in the selected receive group and on the same timeslot as the selected channel. If are using a repeater or a duplex hotspot and you decide to use receive groups, you will need at least two. One for each time slot.

2. If you turn on the dual-channel display (**press and hold P1**) you could use your TS1 monitor channel and your TS2 monitor channel at the same time.

3. It is very easy to get confused and attempt to call a station on the incorrect talk group. You can hear them, but they will not hear you unless you change channels and transmit on the one with the talk group the other station used.

4. You can monitor a dynamic or auto-static talk group, but only if you or someone else keys it up.

5. If you are listening to someone who is using a talk group, you will not hear a call on a different talk group unless you are monitoring both time slots. The repeater can only forward one connection (per time slot) at a time.

Use the CPS to make some receive group lists

Click Digital > receive group Call List. Double click a blank line to add a new receive group, or an existing entry if you want to edit.

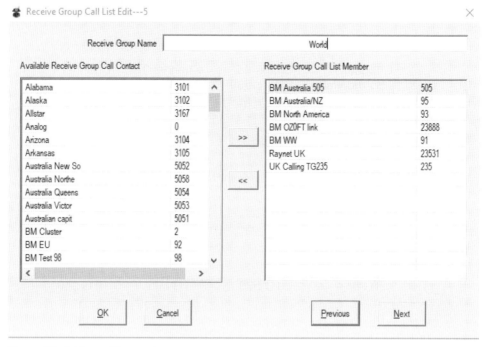

Figure 19: Add or edit a receive group

On the left is a list of all of your talk groups and any private call contacts you put into the talk group list. On the right is a list of the talk groups you have added to the receive group list.

Receive group name - add a name for your receive group list

Select a talk group and use the >> button to move it into the receive group

Select a talk group and use the << button to move it out of the receive group

Press OK to save and exit

Now any channel that has the 'World' receive group list selected will hear all of the selected channels, provided that they are static or have been keyed up.

TIP: add repeater or duplex hotspot static talk groups that are on the same time slot as the channel time slot.

Implementing Andrew's cunning plan

This is what I ended up doing. You may have excellent reasons for doing it completely differently, and that is fine. Do whatever works best for you.

1. Create a receive group (as above) and populate it with the repeater or duplex hotspot's TS1 static talk groups. Name the receive group something like "Local RX group," or whatever you feel is appropriate for the selected talk groups.

2. Create another receive group (as above) and populate it with the repeater or duplex hotspot's TS2 static talk groups. Name the receive group something like "World RX group," or whatever you feel is appropriate for the selected talk groups.

3. Click **Common Setting > Channel**. Select the TS1 channel you want to duplicate and copy it to the clipboard using (Ctrl-C or right click Copy). I copied the channel on my hotspot that is linked to ZL national TG 530. Select a blank channel position and paste (Ctrl-V or right click Paste) the channel.

4. Double click the new channel and change its name to 'TS1 Monitor.'

5. Set the **receive group list** dropdown to 'Local RX group' or whatever your TS1 receive group is called.

6. Click **OK** to save and exit.

7. Select the TS2 channel you want to duplicate and copy it to the clipboard using (Ctrl-C or right click Copy). I copied the channel on my hotspot that is linked to BM Australia TG 505. Select a blank channel position and paste (Ctrl-V or right click Paste) the channel.

8. Double click the new channel and change its name to 'TS2 Monitor.'

9. Set the **receive group list** dropdown to 'World RX group' or whatever your TS2 receive group is called.

10. Click **OK** to save and exit.

11. Go through the other channels for this particular hotspot or repeater and set the **receive group list** dropdown to **None**, for all of the channels except the two monitor channels.

12. Cool! You now have two receive groups and two new channels. Add the two new channels to the zone for the Repeater or hotspot using the CPS.

13. Finally, save the CPS and send the new data to the radio.

What has this achieved?

If you select TS1 Monitor on the radio, you will be able to hear all the calls from the static talk groups in the TS1 receive group. To call someone who was on the receive group, you must change to the channel containing the Talk Group they were using.

If you select TS2 Monitor on the radio, you will be able to hear all the calls from the static talk groups in the TS2 receive group. To call someone who was on the receive group, you must change to the channel containing the Talk Group they were using.

You will also be able to hear and make calls on the talk group associated with the channel. If you select any other channel, you will only hear calls on that talk group.

Save and upload

Save the code plug using the **Save** icon or **File Save**.

Connect the radio and upload the file to the radio. See 'Upload and download data from and to the radio (AnyTone/Radioddity)' on page 31.

SUMMARY

Any time you want to add another talk group, you need to add a channel for it, and then add the channel to the zone. You might have already loaded all the talk groups for your preferred DMR network, in which case you only need to add a channel, and then add the channel to the zone. I am sure you will have hours of fun adding new channels and zones. The whole process is rather frustrating, but eventually, you will end up with the channels you want, containing the talk groups that you are interested in. Although there are thousands of talk groups, only a few are busy.

Using the radio. There is another chapter for that, but basically… Turn on the radio, use the **Up-Down** buttons to select the zone that you want. Each zone includes the channels relating to the repeater or hotspot. Turn the channel knob to select the talk group. Listen to check that the frequency is clear and make a call. Most people do not use CQ on DMR. Say your callsign, "listening for any station on talk group xxx."

PROGRAMMING THE RADODDITY/ANYTONE FOR FM

Programming for FM simplex and repeaters has been covered already, starting on page 51. It is much easier than programming DMR channels. For a start, you don't need to worry about talk groups or hotspots. You only need one FM channel for an FM repeater. It still has to be included in a zone.

I established a zone for all UHF FM repeaters in New Zealand, and a second zone for our 'National System' of linked FM repeaters. I placed the standard FM simplex frequencies in the same zone as the repeaters. My radio is not a dual bander, so I have not included the VHF repeaters, but of course, you can add them into the same zone if you have a dual-band radio.

Programming your TYT radio

The programming options for the TYT radios are a bit different, but the basic functions are the same. This chapter covers programming instructions for the TYT MD-UV380, TYT MD-UV380, and similar models. The previous chapter covered programming instructions for the AnyTone AT-D878UV, the Radioddity GD-AT10G and other radios based on the D878U template. Note that the TYT MD-UV380 is not the same radio as the TYT MD-380. The CPS software for the MD-380 and MD-390 is similar but much less functional.

Start the radio programming software that you downloaded from the manufacturer's website earlier. Most manufactures and websites refer to CPS (Customer Programming Software). It is just another name for the radio configuration software, although how you "program a customer" is unclear.

Start your radio and connect the programming cable between it and the computer. The computer should recognise the device with the usual 'ping.' There is usually no need to load a device driver or select a COM port.

AFTER MAKING CHANGES IN THE CPS PROGRAM

Of course, you must save the edited 'code plug' back to your radio before the changes you have made in the CPS will take effect. Always save a backup copy of the new configuration on your PC as well. Click the **disk icon** or use **File > Save** or **File > Save As**. I called my file **MD_UV380 TYT.rdt**. You could save different versions if you wanted to return the radio to a different configuration.

IT'S ALL TOO TECHNICAL WHAT SHOULD I DO?

All this uploading and downloading files and fiddling with Excel spreadsheets seems way too difficult. Is there an easier way? Sadly, the answer is "no." Unfortunately, DMR was never meant for amateur radio. The configuration of commercial radios is done by the "Technical Boffins." But in the amateur radio world, we have to do it ourselves. That's why we are interested in amateur radio, right? Although it seems challenging, you will soon find that it is not too bad and learn something about spreadsheets along the way.

Once all the Channels, Talk Groups, and Zones have been loaded, using the radio is relatively easy. You only need to set up the radio once. It is the same experience as when you had to add a heap of repeaters to your FM handheld or mobile.

It is possible to program some radios using the buttons on the radio keypad. But you still have to go through the same steps, creating contacts, zones, and channels, and it is extremely slow and tedious doing it that way. There are two other options. Ask someone else to do it or ask a local DMR operator for a 'code plug.'

This is what we are going to do next.

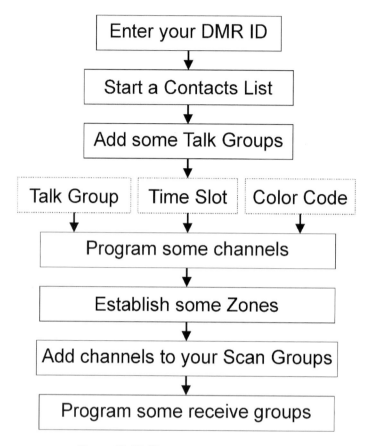

Figure 20: DMR programming stages

ENTER YOUR DMR ID

This only needs to be done once unless you replace the code plug with one from another source. If you do that, you **must** change the DMR ID before using the radio.

On the TYT MD-UV380 CPS, it is under **General Setting**. Fill in the **Radio Name** and **DMR ID** text boxes. The DMR ID is your ID number that you got from RadioID.net back on page 14. The radio name is usually your **callsign − name**, or just your **callsign**, it doesn't matter.

The **Intro Screen** section further down the page allows you to enter some characters such as your name and callsign that will be displayed instead of the standard image when the radio boots up. But I rather like the snazzy TYT logo.

START A CONTACT LIST

If you downloaded a contact list from an Internet site such as RadioID.net or AmateurRadio.digital, you can import it into the CPS software and write it to the radio. See page 24 for the downloading details.

Selecting the TG checkbox adds the full set of Brandmeister talk groups to your contact list. I suggest that you check the box if you will be using the Brandmeister DMR network and don't check the box if you will not be using the Brandmeister DMR network.

The AnyTone/Radioddity radios store separate contact and talk group lists, but unless you do a firmware upgrade, the TYT Digital Contact List contains both contacts and talk groups.

Contacts include the callsign name and other details of other registered DMR users. You can elect to add contacts yourself making a small contact list containing the callsign and details of people you call often. Or for a small annual fee, you can download a contact list from online sources such as RadioID.net. The standard contact list holds 10,000 entries. The optional CSV contact list holds 120,000 contacts.

Talk groups are used to connect your radio to other users with a common interest. Such as the Worldwide talk group on TG 91, or the VK/ZL talk group on TG 95.

To view the contact list, click **Digital Contact**, or **Edit > Digital Contact**. You can edit the contacts by clicking in the text boxes. Or click **Add** to add a new contact.

TIP: If you download a contact list for the TYT MD-UV390 (GPS) from RadioID.net, it will be fine for the MD-UV380 or UV390 with or without GPS. But it will probably need editing. I tried downloads from RadioID.net and AmateurRadio.digital and in both cases the .csv file had to be changed before it would import into the TYT CPS program or EditCP. I find this strange since the suppliers of the data know the format that is required.

The TYT CPS wants a .csv file with columns for **Contact Name, Call Type** (1 for a group call ID or 2 for a private call ID), **Call ID** (the talk group number or contact DMR ID), and **Call Receive Tone** (almost always 0).

TYT firmware upgrade TYT MD-UV380 / MD-UV390

The advanced firmware adds a 'CSV contacts' list capable of storing up to 120,000 contacts. The procedure for carrying out a firmware update starts on page 114. Although updating firmware is always a bit scary, I recommend that you do the upgrade. You get much better information about the stations that you hear. This modification makes the TYT radios more like the AnyTone/Radioddity/BTECH radios. With a large 'CSV' contact list onboard, we can trim the old contact list down so that it only contains the DMR talk groups and any private call IDs that you want to talk with.

Excel Exercise

You should be able to do this with any spreadsheet program that can open a .csv file, such as Excel, Open Office etc. Open the .csv file that you downloaded and save a copy to mess around with. Call it 'MD-380_Contacts.csv' or something similar.

1. The first spreadsheet column must contain the contact and talk group names. If it does, click in cell B1 and insert three new columns. Then skip to item 3.

2. If it does not, click in cell A1 and insert four new columns. Now click the column letter that contains the contact and talk group names. That should select all the cells in the column. Use CTRL+C (or right click and select Copy) to copy the records to the clipboard. Then click in cell A1. Use CTRL+V (or right click and select Paste). Now column A should contain all the contact and talk group names. You can delete the old name column by clicking the column header, right-click, and selecting 'Delete.'

3. The third column needs to contain the TG number and DMR IDs for private calls. It may be labelled 'Call ID.' Click the column letter that contains the TG numbers and DMR IDs. It should select all the cells in that column. Use CTRL+C (or right click and select Copy) to copy the records to the clipboard. Then click in cell C1. Use CTRL+V (or right click and select Paste). Now column C should contain all the contact and talk group names. You can delete the old TG number column. Click the column header, right-click and select 'Delete.'

4. The second column needs to contain 1s for group calls and 2s for private calls. There are three ways to do this. Click in cell B1.

    a. If you have a column X with 'Group' and 'Private' in it. Enter this formula into cell B1.

       =IF(X1="Group",1,2)

    b. If you have a column Y with 'Group Call' and 'Private Call' in it. Enter this formula into cell B1.

       =IF(Y1="Group Call",1,2)

    c. If you do not have a 'Call Type' column, enter this formula into cell B1. =IF(C1>999999,2,1)

    d. Now we are going to copy the formula down the column. Click in cell B1 and use CTRL+C (or right click and select Copy). Hold down the CTRL+Shift+⇓ keys. It should select all cells in the column. Now use CTRL+V (or right click and select Paste) to copy the formula down. When you save and reopen the file the formulas will be gone but the '1' and '2' numbers will remain.

    e.   If cell A1 contains a title, then type "Call type" in cell B1 and "Call ID" in cell C1. The header row is not imported into the CPS.

5.   The 4th column (Call Receive Tone) can be left empty.

TIP: The CPS ignores the first line of the spreadsheet when the file is imported. If the file does not have titles in row 1, you will lose the first talk group. Anything in columns higher than D is ignored.

Important

If you have done the firmware upgrade, you should save up to 120,000 contacts and no talk groups into a file called 'MD-UV380 CSV contacts.csv' or similar. Your talk groups and any private call contacts that you want to make a private call to are saved into the standard contact list. Call that 'MD-UV380_Contacts.csv' or similar. The CSV contact list provides the data that is displayed when you hear a station. The standard Contact list provides the talk groups and the private ID numbers of stations that you plan to transmit to.

If you have not done the firmware upgrade, you will save your talk groups and any private call contacts that you want to make a private call to, and as many other contacts as will fit, into the standard contact list. Call it 'MD-UV380 Contacts.csv' or something similar. The contact list provides the talk group and the private ID numbers of stations that you plan to transmit to, and a limited amount of display data for the few thousand stations in the list.

TIP: You will probably only talk to a few of the 206,000 DMR users in the world, but it makes no difference to the radio whether the contact list has 10 entries or 10,000, so you might as well download as many contacts as will fit in the radio. You certainly want all the contacts from your country or state. It depends a lot on whether you plan to use your DMR for worldwide contacts, nationwide contacts, or local contacts.

Tidy up the list

Before you upload the .csv file to the CPS program you should make a few changes.

If you are using the Brandmeister network, you will have included the Brandmeister talk groups in your downloaded .csv file. There are dozens that you will never use. I deleted all the non-English speaking groups, the Russian, Swedish, Finnish, Romanian, Croatian chat groups, and a heap of others.

If you are not using the Brandmeister network, you will have no talk groups and will need room to add some. I cover how to add the talk groups in the next section, 'Add some talk groups.'

The talk groups can be at the top of the contact list, or at the bottom, or even mixed in with the private call contacts, (although that is a bit horrible).

Save and upload

Save the digital contacts .csv file as 'MD-380_Contacts.csv' and import the file into the CPS software using Digital Contact > Import.'

If you did the firmware update, save the .csv file as 'MD-380_CSV Contacts.csv' and import the file into the CPS software using the menu bar at the top of the CPS page, Program > Write Contacts > Import. Select the CSV Contacts file and Open it.

Save the code plug using the Save icon or File Save.

Connect the radio and upload the file to the radio. See 'Upload and download data from and to the radio (TYT MD-UV380/390)' on page 31.

ADD SOME TALK GROUPS

Before you add talk groups (TG) you need to decide which DMR networks you will be using. Your local repeater will be connected to a DMR network, but other repeaters you can access may be connected to different DMR networks. You can set your hotspot to work on any network that you have subscribed to.

My radio will not accept two talk groups with the same TG number, but there are surprisingly few duplicates. When the big networks use the same TG number, it often means that the talk groups are linked, or that they serve a similar region. So, you only need one TG number anyway. For example, TG 530 on Brandmeister is 'New Zealand.' On DMR+ it is 'ZL national.' They are not linked but the name is close enough that the same TG number can be used on both networks. You can make the name anything you want. The network only uses the number.

Whatever network you are using, TG 99 TS1 is a good choice for digital simplex operation.

Remember to add a talk group for TG 9 (Local). It is used for local calls on a repeater and for calls to a DMR reflector if you are using a hotspot on the DMR+ network.

Brandmeister DMR network

If you are using the Brandmeister network, you will probably have downloaded the Brandmeister talk groups when you downloaded the contact list. If you don't want any more TGs you can skip to the 'programming some channels' section on page 84.

Finding talk group lists

You can download a list of talk groups or build your own. Even if you do download a list, you will probably have to edit it to make it acceptable for your CPS program to import. See talk group lists on page 25.

Most US stations use DMR repeaters which are part of a regional or Statewide network, or they connect their hotspots to the Brandmeister network.

UK hams have access to; FreeDMR, Phoenix UK, Northern DMR Cluster (NDC), SALOP Cluster, South West Cluster (SWC), DV Scotland, FreeSTAR and possibly others. They are regional networks that are permanently linked to DMR+ IPSC2 servers. Some talk groups may be linked across one or several networks. Or you can link your hotspot directly to DMR+ or Bransmeister.

FreeDMR, HB, FD, and TGIF are separate DMR networks that are not linked to DMR+, although they may have some common talk groups.

The 14 New Zealand ZL-TRBO (DMR-MARC) network repeaters are connected to the 'New Zealand' IPSC2 DMR+ server.

TIP: My hotspot is primarily set up for the Brandmeister network, but I need the DMR+ network to participate in the digital voice net on the ZL-TRBO repeaters, so I have both networks set up on my duplex hotspot, [and TGIF].

Your local amateur radio club or DMR operators may be able to provide you with a suitable talk group .csv file.

Option 1: Use a spreadsheet to add talk groups to the contact list

Once you have found a talk group list you have to add it to your contact list. I believe it is best to put the group call talk groups before the private call DMR IDs, just in case the list is longer than the capacity of your radio. It is also a good idea to delete any talk groups that you are not going to use.

If you have done the firmware upgrade, you will probably use the standard contact list for talk groups and have all of the other contacts in the CSV contact list. If you have not done the firmware upgrade, the standard contact list will contain your talk and private ID contacts.

Excel Exercise

Find a talk group list and copy it onto your clipboard using CTRL-C (or right click and Copy). Open a blank spreadsheet and paste in the TG list. CTRL-V (or right click and Paste).

Rearrange the list the same way we did in the 'Start a contact list' Excel Exercise.

- Column A contains the talk group names.

- Column B contains 1 for all talk groups (group calls) and 2 for all private IDs. The only private calls in your talk group list will probably be the 'Parrot' or 'Echo' talk group, and maybe a few friends.

- Column C contains the talk group (TG) number

The TG number does not have to be consecutive. But it must be the same as the network provider uses, or you will end up at the same party but in the wrong room. The CPS will not upload the file if there are duplicate TG numbers. For a private call, enter the DMR ID of the station you want to include.

- Column D (Call Receive Tone) can be left empty

Now, go through and delete the talk groups that you don't want. They are just taking up space.

Count the number of talk groups you have left.

Save the file as a Talk Groups.csv file, but don't close the file. This step is optional, but it can save rework.

Open your Contacts.csv file, in a spreadsheet such as Excel. My file is called 'MD-380_Contacts.csv.'

- Insert enough blank lines at the top of the contact list to fit in the talk groups.

- Copy and paste the talk groups into the top of your Contacts.csv file

- Close the Talk Groups.csv file

Save and upload

Save and close the .csv file as 'MD-380_Contacts.csv' and import the file into the CPS software using Digital Contact > Import.'

Save the code plug using the Save icon or File Save.

Connect the radio and upload the file to the radio. See 'Upload and download data from and to the radio (TYT MD-UV380/390)' on page 31.

	A	B	C	D	E
1	Contact Name	Call Type	Call ID	Call Receive Tone	
2	Cluster	1	2	0	
3	Regional	1	8	0	
4	Local	2	9	0	
5	World-wide	1	91	0	
6	Europe	1	92	0	
7	North America	1	93	0	
8	Asia Middle East	1	94	0	
9	Australia New Ze	1	95	0	
1701	Lester Mcgrath -	2	3303028	0	
1702	Hector Roman-Cru	2	3303021	0	
1703	Julio Cantres Ri	2	3300045	0	

Figure 21: Typical TYT contact list spreadsheet

Option 2: Use the CPS program to add talk groups to the contact list

If you do not want to use Excel or another program to edit .csv files, you can add, edit, and delete talk groups directly in the CPS program. It is quicker if you are only making minor changes, but not if you are doing bulk changes.

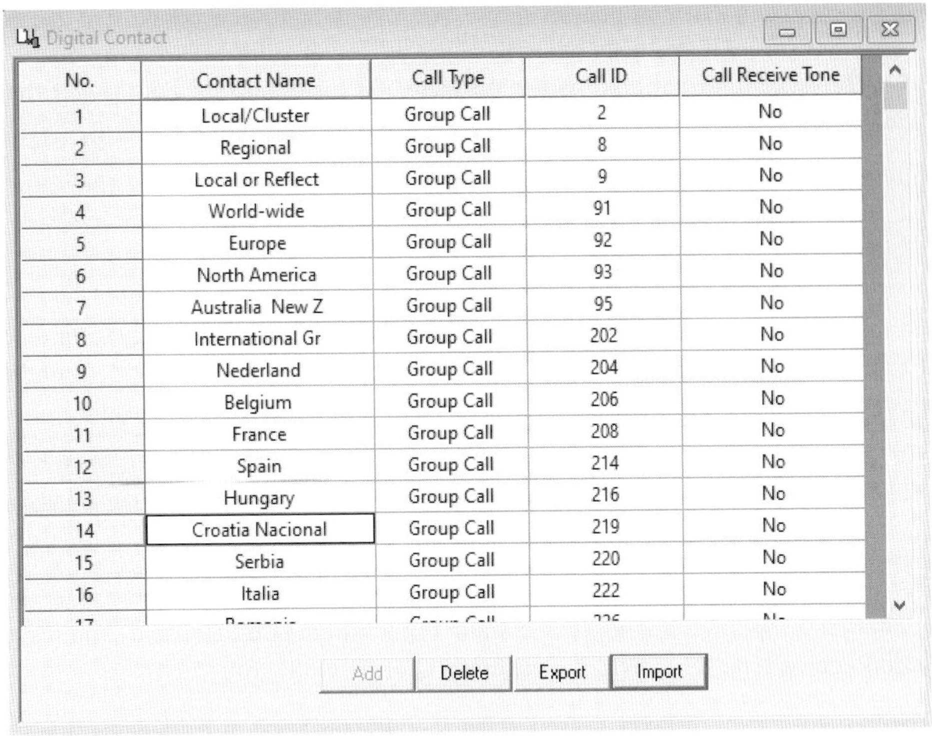

No.	Contact Name	Call Type	Call ID	Call Receive Tone
1	Local/Cluster	Group Call	2	No
2	Regional	Group Call	8	No
3	Local or Reflect	Group Call	9	No
4	World-wide	Group Call	91	No
5	Europe	Group Call	92	No
6	North America	Group Call	93	No
7	Australia New Z	Group Call	95	No
8	International Gr	Group Call	202	No
9	Nederland	Group Call	204	No
10	Belgium	Group Call	206	No
11	France	Group Call	208	No
12	Spain	Group Call	214	No
13	Hungary	Group Call	216	No
14	Croatia Nacional	Group Call	219	No
15	Serbia	Group Call	220	No
16	Italia	Group Call	222	No
17	Romania	Group Call	226	No

Add | Delete | Export | Import

Figure 22: Brandmeister talk groups in the TYT CPS

Add or edit a talk group in the CPS

At a minimum, you need to create talk groups for

- 'Local' on TG 9. TS1 is used when you want to talk with someone over the local repeater, but don't want the call to be broadcast to other linked repeaters. TG 9 TS2 is used to talk on a repeater that is connected to a reflector.

- 'Regional' on TG 8 TS1. TG 8 is often linked to a regional talk group that is only available within your immediate DMR network. On my hotspot, I use TG 8 TS2 to talk on the DMR+ reflectors. Pi-Star translates it to TG 9 on DMR+.

- 'Parrot' or 'Echo' on TG 9990 (depending on the DMR network you will be using)

- 'Disconnect' on TG 4000, or TG 400 on Phoenix UK, if you plan on adding any dynamic talk groups

- and 'Simplex' on TG 99

You also need to add the static talk groups that are available on the repeater. In fact, for all the repeaters that you plan to access. All repeaters in your region that are on the same network will probably have the same or similar static talk groups. See the 'Finding repeater information' chapter on page 133 for details on how to get that information.

Finally, you can add the dynamic talk groups that you are interested in. They must match the DMR network for the repeater, or hotspot that you are planning to access.

Let's start by adding 'Local', or 'LCL' on TG 9.

Open the TYT CPS **Digital Contact** list. Check that TG 9 does not already exist then edit a line at the top of the list or click **Add** to add the talk group to the bottom of the list.

TIP: If the Add button is greyed out, it means that the contact list is full. Just overwrite one of the private ID contacts. The odds that you delete someone you end up working are about 200,000 to 1.

Enter 'LCL TG 9,' or 'Local TG 9,' or 'Local/Ref TG 9' in the **Contact Name** field, Set the **Call Type** dropdown box to 'Group Call.' Put 9 into the **Call ID field**, and leave the **Call Receive tone** set to 'No.' Yipee, that's your first talk group!

The process for Parrot or Echo is similar. Click on the next line or **Add** a new line. In the **Contact Name** field, enter 'Parrot' for the Brandmeister network or 'Echo' if the repeater is linked to the DMR+ network. Enter 9990 in the **Call ID** field. This time the **Call Type** dropdown box should be set to 'Private Call' because you don't want half the world listening to your test call. Leave the **Call Receive Tone** set to 'No.' Howzat, that's your 2nd talk group

Let's enter the Brandmeister Worldwide talk group TG 91. Click on the next line or **Add** a new line. In the **Contact Name** field, enter 'WW TG 91.' Enter 91 in the **Call ID** field. Leave the **Call Type** dropdown box set to 'Group Call.' Leave the **Call Receive tone** set to 'No.' Bazinga, that's your 3rd talk group!

We will do one more and then you're on your own. This time we will enter a Private Call contact. First, check that they are not already on the contact list. You cannot have two contacts with the same ID number.

Click on the next line or **Add** a new line. Enter the station's name or callsign in the **Contact Name** field. It might be another ham, or a club callsign etc. Enter their DMR ID in the **Call ID** field.

If they have been active on DMR for a while, their name and ID might be in your CSV contact list (if activated). Or you can look it up by entering their callsign on RadioID.net. Or read it off the display on your radio during a contact.

The Call Type dropdown box should already be set to 'Private Call.' Leave the Call Receive tone set to 'No.' That's your 4th talk group completed!

Save and upload

Save the code plug using the Save icon or File Save.

Connect the radio and upload the file to the radio. See 'Upload and download data from and to the radio (TYT MD-UV380/390)' on page 31.

Option 3: Use the radio keypad to do a quick add or edit

If you don't have a computer and programming cable, or you just want to add one more talk group or make a quick edit. You can "do things the hard way."

TIP: You enter text using the ABC layout on the keyboard. One press highlights A, a second press highlights B etc. the same as using numeric the keyboard on an old phone. The red - - key is backspace delete. The 0 key is a space. The # key changes the entry mode from letters, to numbers, to characters. This is indicated in the top right of the entry screen. In EN mode pressing a key repeatedly cycles through the three lowercase options, the number, and then the three uppercase options.

- Pressing the red Back (- -) key repeatedly backs you up through the menu structure until you end up back on the main screen.

- Use the up and down arrows to move through the list and click Menu (-) to make each selection.

To add a talk group

- Click the Menu (-) button, and then click it again to select Contacts

- Use the down button to select New Contact. Click Menu (-)

- Select Group Call if you are adding a talk group, or Private Call if you are adding a private call ID.

- Enter the talk group number or the private call ID and click Menu (-) to confirm.

- Enter the talk group name or a contact name and Confirm

- Select 'Tone Off' and Confirm

- That's it. You cannot add any other details from the radio keyboard.

- Note, the radio will not let you save an entry if the talk group ID already exists. If the talk group already exists, you can allocate it to a channel in the next section.

To edit a talk group using the radio keypad

- You can display the talk group number, but you cannot edit a talk group from the radio keyboard.

Save and download

The changes you have made are already in the radio, but if you usually use the CPS, they will not be in that. Next time you open the CPS make sure that you download the current code plug from the radio to the PC before making any changes in the CPS. Otherwise, when you upload back to the radio the changes you entered directly on the radio will be lost. It is always a good idea to save the CPS configuration .rdt file every time.

Connect the radio and upload the file to the radio. See 'Upload and download data from and to the radio (TYT MD-UV380/390)' on page 31.

PROGRAM SOME CHANNELS

You can program new channels using the spreadsheet method, directly in the CPS, or on the radio. For the TYT radios, I prefer the spreadsheet method because you can arrange the channels in a logical order. With the other two methods, you can only add to the bottom of the list. Note that if you change the order of the channels, it will break the relationship with the associated talk groups, and you will have to reset them in the CPS. The CPS does not show channels on a list. It creates a separate channel page for each channel. This is a bit clunky. You cannot rearrange the order of the channels or any of the other lists such as the contacts, zones, scan lists, or RX group call lists, in the CPS.

You will need a channel for every talk group that you want to use on every repeater that you want to use. This means that you have to create many channels for each repeater frequency pair. I don't know why the CPS software makes this such an ordeal, I would have thought that you could create one RF channel and then allocate multiple talk groups to it... but unfortunately, it does not work that way. You also need to allocate the correct time slot and colour code to each channel. The colour code will be the same for all channels on a particular repeater, but the time slot has to suit the talk group.

Finding the right talk group,

If you have the DMR channel programmed and you can see transmissions on the RSSI (receive signal strength indicator), but you cannot hear anything.

You may be listening to a different talk group, or you might have the wrong time slot or colour code selected.

Use the 'monitor' to find out the talk group, colour code and time slot for the repeater. You can turn on the monitor by pressing the button below the PTT on the radio. When a call is made the display will show the 'TxContact' which is the talk group number, and the CC (colour code). 'Type' is G for a group call. 'Solt' is not a misspelling of the time slot. It indicates a '2' if the radio is listening to a repeater or a duplex hotspot and the frequencies are different. It is not shown at all if the radio is listening to a simplex hotspot or the frequencies are the same.

Colour Code (CC)

The colour code is used to identify the output of a specific repeater, much like the CTCSS tone on an FM repeater. In the unlikely event that two repeaters near you are using the same frequencies. Using different colour codes can ensure that you only hear the wanted repeater. If in doubt, try CC01. I have been told that all VK and ZL repeaters use CC01. If you have set the colour code incorrectly you will not be able to hear the repeater. Unless you have the digital monitor enabled.

Time Slot (TS)

A repeater or duplex hotspot transmits two time slots. Your channel must have the right one or you will not be able to hear or talk on the wanted talk group. A simplex hotspot only transmits on TS2.

Option 1: Use the CPS to make a channel list

I believe that entering channels directly into the CPS is much less confusing than using the spreadsheet entry method. Although it is not a bad idea to enter your channels using the CPS and then rearrange them into a logical order using the spreadsheet method.

You have to create a channel for every talk group that you want to use on the specific repeater or hotspot. The talk groups you select must already be in your talk group list.

A) Before you start

You will need to know the repeater or hotspot's transmit frequency, receive frequency or offset, colour code, DMR network, and static talk groups (if any).

Navigate to the **Channel Information** folder and click to select the first channel name. There should be at least one, even in a new radio. This will pop up

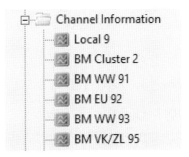

Figure 23: TYT Channels

the edit screen for the first channel. You can begin by editing the first channel and then proceed to **Add** more channels.

B) Program an FM channel

We will start with an FM channel for an analog repeater. You only need one channel for each FM repeater. Skip this step if the radio will only be used for DMR.

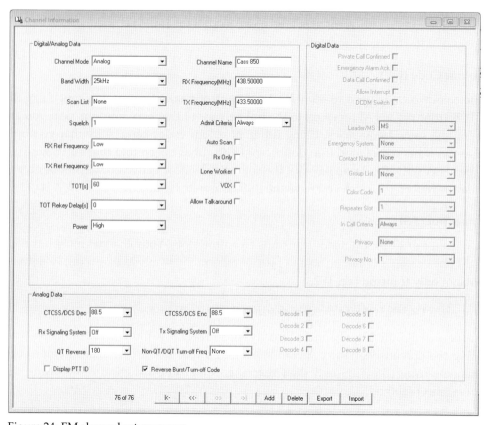

Figure 24: FM channel setup screen

Channel mode – **Analog**

Bandwidth is always **25 kHz** for analog channels

Scan List – select a scan list if you have one set up for FM channels or select **None**

Squelch – usually set at **1**

RX Ref Frequency - **Low**

TX Ref Frequency - **Low**

TOT[s] Time out timer – **60 seconds**

TOT Rekey Delay[s] – 0 seconds

Power – set the transmitter power to a level that ensures you can reliably access the repeater. My radio has three power levels, High, Medium, and Low.

TIP: use the lowest power possible that provides reliable communications to extend battery life. Higher transmit power can dramatically increase battery drain. I would normally set the Transmit Power to High for repeaters.

Channel name – enter a channel name that identifies the repeater. In New Zealand that consists of the repeater name 'Cass Peak' and a code representing the frequency. The repeater transmits on 438.500 so the code is 850.

Receive frequency – set this to the repeater output frequency.

Transmit frequency – set this to the repeater input frequency.

Leave the five checkboxes unchecked.

TIP: The 'Allow Talk around (Simplex)' checkbox sets the transmitter to the stated receiver frequency. It can be used if you want to temporarily transmit on the output frequency of a repeater.

The bottom section is for tone squelch. Most FM repeaters use CTCSS, or no tone squelch. DCS is rare and as far as I know, DTMF is not used anywhere. If your repeater needs a CTCSS tone, set CTCSS/DCS Encode to CTCSS and use the dropdown list to select the correct CTCSS tone for the repeater.

Generally, the repeater will send the CTCSS tone on the output as well, so you can set CTCSS/DCS Decode to CTCSS and use the dropdown list to select the same CTCSS tone for your receiver.

QT reverse (Quiet Talk) leave at 180, not used for amateur radio.

Non-QT/DQT Turn off Freq (Digital Quiet Talk) - None not used for amateur radio

Reverse Burst/Turn off code – This has something to do with eliminating the repeater 'tail' "kerchunk" noise. It turns itself on when you select a CTCSS Encode, so I guess you can leave it selected.

Display PTT ID – unselected not used for amateur radio.

C) Program an FM simplex channel

An FM simplex channel is the same as an FM repeater channel, except the receive and transmit frequency will be the same, and you would not normally use CTCSS tones. Set CTCSS/DCS Encode and CTCSS/DCS Decode to Off.

D) Program a DMR channel

You should program a channel for every talk group that you want to use on each repeater or hotspot.

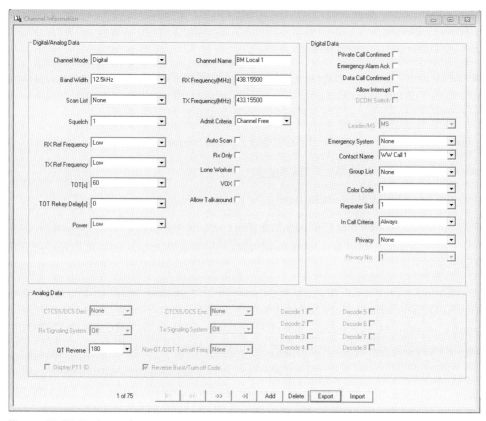

Figure 25: DMR channel setup screen

Channel mode – Digital

Bandwidth is always 12.5 kHz for digital channels

Scan List – select a scan list if you have one set up for digital channels or select None

Squelch – usually set at 1

RX Ref Frequency - Low

TX Ref Frequency - Low

TOT[s] Time out timer – 60 seconds

TOT Rekey Delay[s] – 0 seconds

Power – set the transmitter power to a level that ensures you can reliably access the repeater. My radio has three power levels, High, Medium, and **Low**.

TIP: use the lowest power possible that provides reliable communications to extend battery life. Higher transmit power can dramatically increase battery drain. I would normally set the Transmit Power to High for repeaters and low for hotspots.

Channel name – enter a channel name that identifies the channel. On the TYT radios, **all channels must have a unique name. The zone list will not select the correct channel if you have two (or more) with the same name! (YUCK!)**

I like to include the talk group number. I also tag my DMR+ channels so I know what network I am on. For example, ZL 530 on Brandmeister is not the same as ZL 530 on DMR+.

Brandmeister channel names	DMR+ channel names
ZL1 region 5301	DMR+ UK 235
ZL2 region 5302	DMR+ VK 505
ZL3 region 5303	DMR+ Oceania 5
ZL4 region 5304	DMR+ ZL 530
BM Parrot	DMR+ Monitor
UK Chat 23531	DMR+ Quadnet 320
OZ0FT link 23888	DMR+ TAC EN 153
VK6 hub 505005	DMR+ WWE 13

Receive frequency – set this to the repeater or hotspot output frequency.

Transmit frequency – set this to the repeater or hotspot input frequency.

Admit Criteria – **Channel Free** only allows your radio to transmit if the channel is free. It is OK for a simplex hotspot. The **Color Code** setting only allows the radio to transmit if the received colour code matches your channel setup. I use the Color Code setting because if I use Channel Free the handheld will not transmit on TS1 if the hotspot or repeater is transmitting on TS2, (and vice versa). **Always** allows the radio to transmit anytime you press the PTT, which could interfere with other users, but it is a good option for your TG 4000 disconnect channel.

Leave the five checkboxes unchecked.

All of the information in the 'Digital/Analog Data' section on the left will be the same for every channel you add to the same repeater or hotspot. You can copy and paste the information so that you don't have to type it in for every channel. See 'Program a subsequent channel' on page 91.

The 'Digital Data' section on the right is the most important part of the process.

Leave the five checkboxes at the top of the section unselected. They are not for amateur radio use.

Leader/MS is greyed out and I do not know what it does.

Emergency system – leave at **None**. It is not for amateur radio use.

Contact Name is the most important setting of all. It associates the talk group with the channel. You need one channel for each talk group that you want to use. Select a talk group that matches the channel name. For example, if you called the channel ZL 530, then select the ZL 530 talk group. It gets very confusing if you get this mismatched.

TIP: Do not create channels for talk groups that cannot be accessed over the repeater or hotspot. For example, if you are creating channels for your local repeater on the Southern Wahoo DMR network, and the network is linked to DMR+, there is no point in adding Brandmeister talk groups. If you are creating channels for a Brandmeister repeater, there is no point in adding DMR+ or TGIF talk groups.

Group List. This is where you select a receive group list. I recommend setting up two monitor channels, one to monitor all your talk TS1 groups, and the other to monitor all your TS2 talk groups. They will be associated with your most popular TS1 and TS2 talk groups. If you are using a simplex hotspot, you only need one monitor on TS2. All other channels will have the Group List set to **None**. See receive group lists on page 68.

Colour code – will be the same for all channels on the repeater or hotspot but could be different on another repeater. If in doubt, try **CC1**. You can use the monitor to find out what colour code is in use.

Repeater slot – is the time slot you will transmit on. It should match the requirements of the repeater. The repeater owner will usually specify that local calls be made on one time slot and other calls be made on the other time slot. If you are setting up channels for static talk groups, they must match the time slot on the repeater. If you have a duplex hotspot, you can arrange your own structure and static talk groups. If you have a simplex hotspot all channels will use TS2.

TIP: My hotspot is set up with Brandmeister on TS1 and DMR+ on TS2. Some of the DMR+ talk groups are on TS1, but you can arrange the Pi-star Software to cater for that.

In Call Criteria should be set to **Always** or **Follow Admit Criteria**.

Privacy must be set to **None**. In most countries, amateur radio operators are not allowed to use encryption. And it will not work anyway because nobody else uses it.

Program a subsequent DMR channel

Well done! You have programmed your first DMR channel. The good news is that the second one will be much easier.

Click **Add** at the bottom of the channel information page to create another channel.

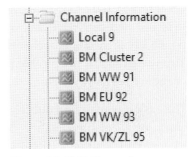

Figure 26: TYT Channels

In the channel list, right click on the channel that you just finished and select 'Copy.'

Find the newly added channel at the bottom of the list of channels. It should be called 'Channel1.' Right-click the new channel and select 'Paste.' This will copy the basic radio information such as the power and frequencies into the new channel.

Now, all you have to do is enter the new **Channel Name,** and the **Contact Name,** then check the **Repeater Slot** is OK for the new channel.

E) Program a digital simplex channel

A digital simplex channel is the same as a DMR repeater channel, except the receive and transmit frequency will be the same. For the **Contact Name,** select your simplex talk group on **TG 99 TS1.** The time slot does not matter for simplex because the radio will transmit on TS1 anyway. But the colour code must match the radio you are calling. You can use the monitor to check what colour code the other station is using.

The other way to make a simplex channel is to make a private call channel. The receive and transmit frequencies will be the same. Select the name of the private call ID of the other radio as the **Contact Name.** Make sure both radios are using the same colour code and time slot. You can do this between two of your radios that have the same DMR ID. Load your DMR ID as the **Contact Name** on both radios.

F) Save and upload

Save the code plug using the **Save** icon or **File Save.**

Connect the radio and upload the file to the radio. See 'Upload and download data from and to the radio (TYT MD-UV380/390)' on page 31.

Option 2: Use a spreadsheet to produce a channel list

I think that it is generally easier to enter new channels using the CPS program. But you can use a spreadsheet to rearrange the channels into a more sensible order. Grouping all the channels for a particular repeater or hotspot together is a good start. It is not necessary, but it can make the programming less confusing. The other possibility is installing a channel list .csv file that has been created by someone else. You will probably not be able to find a channel list online, but your local DMR group or radio club may have one, or another DMR user.

The easiest way to find out what the CPS program will accept is to enter one channel entry, or all of the channels for one repeater, using the CPS and then export the 'channel' file and enter the remaining channels, or rearrange the channel order in a spreadsheet. Navigate to the **Channel Information** folder and click to select the first channel name. There should be at least one channel, even on a new radio. This will pop up the edit screen for the first channel. Click the Export button at the bottom of the page to save the current channels as a .csv file. I made a CPS TYT folder for all files relating to the TYT CPS. Save the file as 'UV380_channels' or something similar.

TIP: the TYT CPS has the rather annoying habit of adding .csv to any filename. That is fine when you export a file for the first time, but if you upload the channel list, edit it, and then export it again it will become filename.csv.csv. Next time it becomes filename.csv.csv.csv and so on. You have to go into Windows Explorer to edit the file name and get rid of the additional filename extensions.

Now open your spreadsheet and open the new UV380_channels.csv file.

My TYT CPS is expecting a file with fifty-one named columns. Most of them can simply be copied down the list.

When you have finished editing the .csv file, save it on your PC and then import it back into the CPS program using the **Import** button on any channel information page.

	Channel Mode	Channel Name	RX Frequency(MHz)	TX Frequency(MHz)	Band Width	Scan List	Squelch
35	2	DMR+ UK 235	438.155	433.155	0	0	1
36	2	DMR+ VK 505	438.155	433.155	0	0	1
37	2	DMR+ Oceania 5	438.155	433.155	0	0	1
38	2	DMR+ ZL 530	438.155	433.155	0	0	1
39	2	DMR+ Monitor	438.155	433.155	0	0	1
40	2	DMR+ Quadnet 320	438.155	433.155	0	0	1
41	2	DMR+ TAC EN 153	438.155	433.155	0	0	1
42	2	DMR+ WWE 13	438.155	433.155	0	0	1

Figure 27: The most important columns in the Channel .csv file

If you have channels for several repeaters that are on the same network, such as the 14 repeaters in the ZL-TRBO network, you will see the same pattern of channels for each repeater. For each repeater or hotspot, you need a channel for each of the static talk groups, dynamic talk groups you might use, local TG 9, disconnect TG 4000, and Parrot or Echo TG 9990.

You can select and drag many of the entries down the list as the entries are the same on each line. Some settings cannot be changed in the CPS and are probably not relevant to the radio model.

Channel mode is 2 for digital and 1 for analog

Channel Name is usually a combination of the talk group name and the channel name. Whatever is going to be easy to remember. See the image above. If the repeater or hotspot is only connected to one DMR network and you only have that repeater or hotspot in the zone, you could omit the 'DMR+' and just put the talk group name. I like to add the talk group number as well. You can choose any name that will help you to remember what talk group you are on. For example, 'EME group,' 'Statewide 3121' if you live in Kentucky, or 'Chat 1' if you live in the UK. **Note that all channels must have a unique name. The zone list will not select the correct channel if you have two (or more) channels with the same name! (TYT bad software development).**

RX frequency (MHz) is the DMR repeater or hotspot output frequency

TX frequency (MHz) is the DMR repeater or hotspot input frequency

Bandwidth is always **0** (12.5 kHz) for DMR and **2** (25 kHz) for FM

Scan List is **0** for None, **1** for scan list 1 etc.

Squelch 1 to 5. Set it to **1** for digital channels. You won't hear anything on a digital channel unless the channel can be decoded, and the talk group colour code and time slot are correct.

The RX Ref and TX Ref frequency are there to create an offset if your hotspot is a little off frequency. It is usually better to correct the hotspot using the Ref offset in the Pi-Star software than to mess around with this offset. Leave them at **0**.

The time out timer (TOT) controls how long the radio will continue to transmit if you wedge the radio down the car seat or the sofa. I leave my radio set to the default 60 seconds. I have seen a comment online that you should set it to 180 seconds because that matches the TOT on most repeaters and hotspots. You should definitely NOT set it to 'infinite.' The numbers in the TOT column relate to the TOT times in the dropdown list on the CPS in 15-second steps.

0 = infinite, 1 = 15, 2 = 30, 3= 45, 4 =60 and so on up to 37 = 555 seconds.

TOT Rekey delay sets a delay after a TOT time out before the radio will transmit again. It can be set from 0 to 255 seconds in 1-second steps. The default is 0 seconds.

Like the TOT setting, the TOT Re-key delay is intended to limit the interference from a "stuck on" mobile.

Power – to save battery power use as little as possible that ensures reliable communications. I recommend 0 (Low) for a hotspot at your home and 2 (High) for repeaters. Guess what power level 1 is!

Admit criteria. 1 = 'Channel Free,' 2 = 'CTCSS' (for FM), 0 = 'Always' (not recommended because you could transmit over another user.

Auto Scan, RX only, Lone Worker, VOX, and Allow Work Around, are usually 0.

Send GPS. My MD-UV380 does not have GPS so this is 0. If your radio has GPS this number may be higher.

Receive GPS info is not a CPS option on my radio. So, it is always 0.

Private Call Confirmed, Emergency Alarm Ack, Data Call Confirmed, Allow Interrupt DCDM Switch, and Emergency System are always 0. Leader/MS is 1.

Contact name is not the contact name or the talk group/ private ID number. It is the row number of the selected contact, on the contact list. This method of indexing means that if you change the order of the talk groups on your contact list it will mess up the associations in the channel list. This is lazy programming by the TYT developer. If they had referenced the TG number instead of the row number this would not happen.

Group List is the receive group list number

Color code is the repeater or hotspot's colour code. It will be the same for all channels.

Repeater Slot is the time slot for the talk group on the repeater or hotspot. The same talk group could be on a different time slot on a different repeater or hotspot.

All of the remaining columns are only for FM

CTCSS/DCS Decode – is for FM repeaters only enter the CTCSS Tone number

CTCSS/DCS Encode – is for FM repeaters only enter the CTCSS Tone number

Save and upload

Save the channel list as a .csv file. Then use open the CPS and select **Channel Information** > click on any **Channel** > and click the **Import** button to import the .csv file back into the CPS. After clicking Import, you will probably have to navigate to find the 'UV380_channels.csv' file and click **OK** to start the upload

Save the code plug using the **Save** icon or **File Save**.

You will see the new channels populate the Channel Information list and can step through them to check the details and make sure that is pointing at the wanted talk groups. When you are happy, save the code plug using the **Save** icon or **File Save**.

Then connect the radio and upload the file to the radio. See 'Upload and download data from and to the radio (TYT MD-UV380/390)' on page 31.

Option 3: Add a channel using the buttons on the radio

Sometimes you might want to add a channel using the radio buttons. You might have driven to a new town or want to use a friend's hotspot. The procedure is tedious but possible.

TIP: If a password is required it means that one has been set in the CPS. I don't think that it is wise to set either the PC Programming or Radio program password.

To add a channel, go to **Menu > Utilities > Program Radio >** go to **No.6 Add**, then select Analog CH or **Digital CH**.

The display will say **Enter CH Name, Channel 1**. Use the up ▲ arrow to backspace and enter a new name using the radio keypad buttons. Multiple presses cycle through lowercase, a number, and uppercase. Press the Menu button to **Confirm**.

Use the up ▲ arrow to backspace and the keypad to enter the receiver frequency (the repeater or hotspot output frequency). Press the Menu button to **Confirm**.

Use the up ▲ arrow to backspace and the keypad to enter the transmitter frequency (the repeater or hotspot input frequency). Press the Menu button to **Confirm**.

Use the up ▲ arrow or down ▼ arrow the select the talk group or private call contact. Press the Menu button to **Confirm**.

Use the up ▲ arrow or down ▼ arrow the select the receive group list. Press the Menu button to **Confirm**. Apparently, you cannot choose a 'None' option!

This completes the 'add a new channel' menu structure. There is no way to set the zone for your new channel. Continue to 'Option 4: Edit a channel,' to make other changes to your new channel. You definitely need to set the zone, colour code, and time slot.

Option 4: Edit a channel using the buttons on the radio

The radio will edit the currently selected 'A' channel, (the upper one). So, select the channel that you want to edit starting at number 1 'RX Frequency,' or carry on from the previous section starting with item 4 'Time Out Timer.'

To edit a channel, go to **Menu > Utilities > Program Radio** then carry on with the settings below.

1. RX Frequency. Press **Confirm** if you want to edit the frequency or **Back** to return to the previous menu level. To edit, use the up ▲ arrow to backspace and the keypad to enter the receiver frequency, (the repeater or hotspot output frequency). Press the Menu button to **Confirm**.

2. TX Frequency. Press **Confirm** if you want to edit the frequency or **Back** to return to the previous menu level. To edit, use the up ▲ arrow to backspace and the keypad to enter the transmitter frequency (the repeater or hotspot input frequency). Press the Menu button to **Confirm**.

3. Channel Name. Use the up ▲ arrow to backspace and enter a new name using the radio keypad buttons. Multiple presses cycle through lowercase, a number, and uppercase. Press the Menu button to **Confirm**.

4. Time Out Timer. Set the TOT. The default is 60 seconds, you can use up to 180 seconds. Do not use 'Infinite,' or anything greater than 180 seconds. Press the Menu button to **Confirm**.

5. CTS/DCS. This is for setting the CTCSS tone and cannot be selected if you are editing a digital channel.

6. Add. This is the selection for adding a new channel. So, don't use it when you are editing an existing channel.

7. Color Code. Use the up ▲ arrow or down ▼ arrow to select the Color Code. Press the Menu button to **Confirm**.

8. Time Slot. Use the up ▲ arrow or down ▼ arrow then select the time slot. Press the Menu button to **Confirm**.

9. Vox. Use the up ▲ arrow or down ▼ arrow the select the VOX setting. Press the Menu button to **Confirm**.

10. TX Contact. Use the up ▲ arrow or down ▼ arrow the select the talk group. Press the Menu button to **Confirm**. (If you just added a channel, this will already be selected. Skip to the next item).

11. Group List. Use the up ▲ arrow or down ▼ arrow the select the receive group list. Or you can create a new receive group list. Press the Menu button to **Confirm**. (If you just added a channel, this will already be selected. Skip to the next item).

12. Mic Level. This affects all of your channels, not just the current one. The microphone level must not be too high. It will cause distortion and make your signal difficult to understand. See 'setting your microphone and speaker settings' on page 154. Use the up ▲ arrow or down ▼ arrow the select the Mic Level. Press the Menu button to **Confirm**.

ESTABLISH SOME ZONES

Zones allocate the channels to the channel switch on the radio. You select the zone that contains the channels for a particular repeater, or hotspot, or region, or city. I have a zone for each repeater and one for each of my hotspots. How you set up your zones is up to you. You could have one zone containing all of your channels. But I think that would be confusing.

*TIP: If you add a new channel and you want to be able to transmit on it. You **must** add it into at least one zone. If you don't there is no way to select the channel.*

TALK GROUP ➡ CHANNEL ➡ ZONE

Zones are a bit like the memory groups on an FM radio. They can be named, and you can use them in various ways. If you travel to another city for work and will be using different repeaters you could have a Zone for 'home' and another for 'work.' Or you could have a zone for your local repeaters and a different one for your hotspot. And perhaps a third one for FM repeaters. After you enter the channels that you will use, you can place them into the zones. It means that if your local zone is selected you will only be able to choose channels that are within reach. When you are in another city, you will select zone two and be able to reach the appropriate repeaters.

The TYT MD-UV380 and MD-UV390 radios can have up to 250 Zones with up to 64 channels each. Click Zone Information and the first zone on the list. You can change the name of Zone 1 if you want to. Click Add to add a second zone and so on. We will do that later. The MD-380 and MD-390 only have 16 channels per zone, but this is plenty. If it becomes a problem, simply add another zone.

Use the CPS to create some zones

You must have at least one zone. I currently have ten. I have two for the local DMR repeaters and several for my hotspot.

The TYT radios have two channel lists for each zone. One is for 'Channel A,' the channel that is higher on the radio's screen and the other is for 'Channel B,' the channel that is lower on the radio's screen. I really like this option; it is one of the few things the TYT radio has that the Radioddity/AnyTone CPS doesn't have.

So far, I have zones for; Hotspot BM, Hotspot DMR+, Hotspot TGIF, Christchurch ZL3DVR, Christchurch ZL3DMR, Digital simplex, APRS, FM 'National System', FM repeaters, and FM simplex.

Have a good think about the way you want to set up your zones. It is worth some careful consideration.

A) Add a new zone (or edit an existing zone)

In the CPS under Zone Information, click the top Zone. It may still be called Zone1.

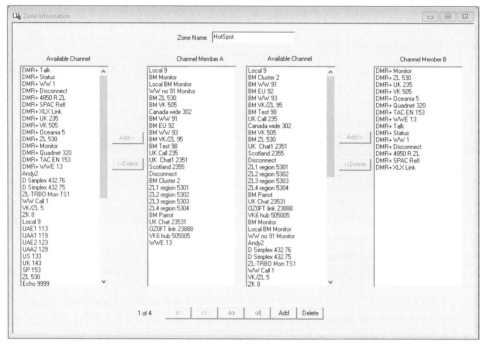

Figure 28: Zones in the TYT CPS

Zone Name – enter a name for your zone. The one shown above is for my hotspot.

Click >> to move a channel into a zone. Click >> again to move the next channel into the zone. Click << to move a channel out of a zone. To avoid display problems after you download a code plug to the radio, **every zone should have at least one channel on side A and one on side B. Especially your first zone!**

TIP: Pick the channels carefully. The order that you put them into the list is the order that they will appear on your radio. You cannot change the order in the list except by taking the channel out of the zone and putting it back in later. You will see that I have placed my monitor channels at the top and my most used channels next. Because I structured my duplex hotspot to have Brandmeister on TS1 and DMR+ on TS2, I thought that it was sensible to put the Brandmeister channels on 'channel A' at the top of the radio display and the DMR+ channels on 'channel B' at the bottom of the radio display. It means that I can monitor the BM channels and the DMR+ channels at the same time. If the zone is for a repeater, I recommend putting a TS1 talk group at the top of the channel A list and a TS2 talk group at the top of the channel B list. That way you can take advantage of the dual time slots on the repeater. A simplex hotspot can only transmit one signal at a time, so it would be better to have a hotspot channel on channel A and a local repeater channel on channel B.

B) Save and upload

Save the code plug using the Save icon or File Save.

Then connect the radio and upload the file to the radio. See 'Upload and download data from and to the radio (TYT MD-UV380/390)' on page 31.

ADD CHANNELS TO YOUR SCAN LIST

You can add any channels to a Scan List, but scanning is better suited to FM. I guess you could scan two or more local repeaters, but since I can configure my hotspot any way I like, I don't bother using scanning on DMR.

*TIP: If the scan stops on an active signal, you can press the green **Menu** button to stop the scan from resuming and stay on that channel.*

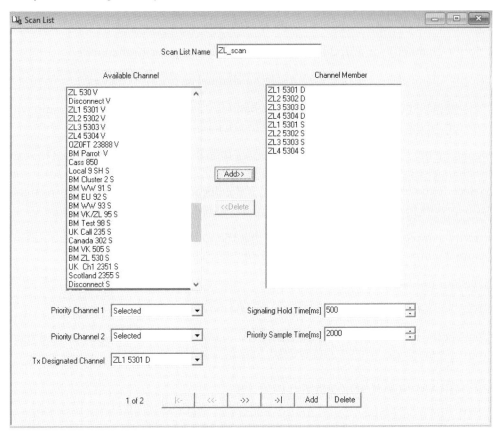

Figure 29: TYT Scan List

The image above shows the same four talk groups on two different repeaters, included in the 'ZL_scan' scan list.

Scan List Name – enter a name for your scan list.

Navigate down the left list to the channels you want in the scan list. They will often be bunched together because you will have created them in a group. This is where prefixing can help. Or just pop into the channel list and make a note of the channel numbers you want. Click >> to move a channel into a scan list. Click >> again to move the next channel into the scan list.

TIP: Select in the right side list and click << if you want to take a channel out of the list.

I have not bothered setting priority channels because the radio is dedicated to DMR and I don't use the scan function. These are functions that you would change from the radio keyboard if you wanted to alter the way the radio scans.

Priority Channel Select – off, or the priority channel 1 selected below, or the priority channel 2 selected below, or both priority channels.

Priority Channel 1 – a dropdown to select a channel as priority channel 1.

Priority Channel 2 – a dropdown to select a channel as priority channel 2.

TX Designated channel – I am not sure what this does. I think it means the radio will transmit on the selected channel and not the channel that the scan stops on.

Signalling hold time [ms] - Sets the amount of time that the radio waits on an FM scan list channel when a carrier signal opens the squelch. This pause allows the radio time to decode the FM DTMF signalling data. If the decoded information is incorrect, the radio reverts to scan. This setting only applies to FM. Leave it at the default.

Priority sample time [ms] sets the time that the radio waits when listening to a call before it jumps away to scan the two priority channels.

Save and upload

Save the code plug using the **Save** icon or **File Save**.

Connect the radio and upload the file to the radio. See 'Upload and download data from and to the radio (TYT MD-UV380/390)' on page 31.

PROGRAM SOME RECEIVE GROUPS

You do not have to use receive groups on the TYT MD-UV380 and MD-UV390 radios. If you do decide to create receive groups they should be planned with care. They work a little like the digital monitor (promiscuous) mode, but they are not as easy to turn off. If you associate a receive group with a channel, you will hear the calls on the talk groups in the receive group as well as the talk group allocated to the channel. This is great to check if there is activity on the repeater. But it can be confusing. For example, you hear someone make a call, but you don't know what talk group to respond on.

You can include dynamic talk groups in the receive list, but you will only hear them if you or another operator transmits to activate the talk group.

TIP: Some older radios require that every channel has a receive group containing at least that channel to hear anything. On most other radios, if you don't select a receive group list when you create the channel, you will only hear the talk group associated with that channel.

Rules governing receive groups

1. The radio will only hear talk groups that are in the selected receive group and on the same timeslot as the selected channel. If you decide to use receive groups, you will probably need at least two. One for each time slot.

2. If you turn on the dual-channel display you could monitor a channel on TS1 and another on TS2, with appropriate receive talk groups.

3. It is very easy to get confused and attempt to call a station on the incorrect talk group. You can hear them, but they will not hear you unless you change to a channel with the talk group the other station used.

4. You can monitor a dynamic or auto-static talk group, but only if you or someone else keys it up.

5. If you are listening to someone, you will not hear a call on the same time slot but a different talk group. The repeater can only forward one connection (per time slot) at a time. The hang time in the Pi-Star software prevents the hotspot from changing to another timeslot during a normal conversation.

My preferred method

It is a personal preference, but I believe that it is a bad idea to allocate a receive group on every channel. I think that it is better to create two additional 'Monitor' channels (per hotspot) and link them to two receive talk groups. One for each time slot. You will be able to switch to the 'TS1 Monitor' channel and hear all the static talk groups on TS1. Or switch to the 'TS2 Monitor' channel and hear all the static talk groups on TS2. When a call comes in that you want to respond to, you switch the radio to the correct talk group and make your call.

Note if you are using a simplex hotspot, everything is on TS2, so you only need one monitor channel.

Use the CPS to make some receive groups

Click **Digital RX Group** Call and select the first receive group. On the left is a list of all of your talk groups and private call contacts. On the right is a list of the talk groups you have added to the receive group list.

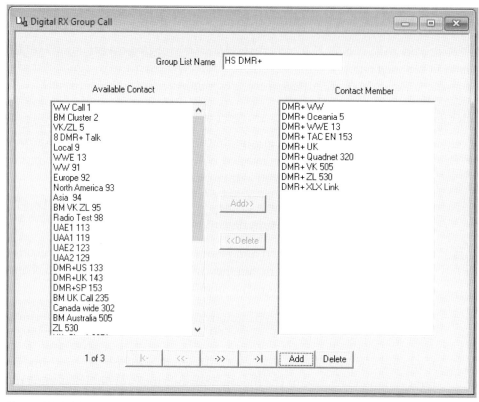

Figure 30: Add or edit a receive group

Receive group name - add a name for your receive group list

Select a talk group and use the >> button to move it into the receive group

Select a talk group and use the << button to move it out of the receive group

TIP: add all of the repeater or duplex hotspot static talk groups that are on the same time slot. Simplex hotspots don't support static talk groups, so just add talk groups that you want to monitor. Keying up a talk will cause it to become auto-static and the previous talk group will become dynamic for 15 minutes. You can monitor both channels if they are included in your receive group.

Andrew's cunning plan

This is what I ended up doing. You may have excellent reasons for doing it completely differently, and that is fine. Do whatever works best for you.

1. Create a receive group (as above) and populate it with the repeater or duplex hotspot's TS1 static talk groups. Name the receive group something like "Local RX group," or whatever you feel is appropriate for the selected talk groups.

2. Create another receive group (as above) and populate it with the repeater or duplex hotspot's TS2 static talk groups. Name the receive group something like "World RX group," or whatever you feel is appropriate for the selected talk groups.

3. Click **Channel Information** and select the TS1 channel you want to duplicate. Copy it to the clipboard using (Ctrl-C or right click Copy). I suggest using the channel with the busiest talk group on the list. Add a blank channel position and paste (Ctrl-V or right click Paste) the channel.

4. Select the new channel and change its name to 'TS1 Monitor.'

5. Set the **Group List** dropdown to 'Local RX group' or whatever your TS1 receive group is called.

6. Repeat the process to create a monitor channel for TS2.

7. Select the TS2 channel you want to duplicate and copy it to the clipboard using (Ctrl-C or right click Copy). I suggest using the channel with the busiest talk group on the list. Select a blank channel position and paste (Ctrl-V or right click Paste) the channel.

8. Select the new channel and change its name to 'TS2 Monitor.'

9. Set the **receive group list** dropdown to 'World RX group' or whatever your TS2 receive group is called.

10. Go through the other channels for this particular hotspot or repeater and set the **Group List** dropdown to **None** for all of the channels except the two monitor channels.

11. Cool! You now have two receive groups and two new channels. Add the two new channels to the zone for the repeater or hotspot.

12. Finally, save the CPS and send the new data to the radio.

What has this achieved?

If you select TS1 Monitor on the radio, you will be able to hear all the calls from the static talk groups in the TS1 receive group. To call someone who was on the receive group, you must change to the channel containing the talk group they were using.

If you select TS2 Monitor on the radio, you will be able to hear all the calls from the static talk groups in the TS2 receive group. To call someone who was on the receive group, you must change to the channel containing the talk group they were using.

You will also be able to hear and make calls on the talk group that is associated with the channel. If you select any other channel, you will only hear calls on that talk group.

Save and upload

Save the code plug using the **Save** icon or **File Save**.

Connect the radio and upload the file to the radio. See 'Upload and download data from and to the radio (TYT MD-UV380/390)' on page 31.

SUMMARY

Any time you want to add another talk group, you need to add a channel for it, and then add the channel to a zone.

TALK GROUP ➡ CHANNEL ➡ ZONE

You might have loaded all the talk groups for your preferred DMR network, in which case you only need to add a channel, and then add the channel to the zone. I am sure you will have hours of fun adding new channels and zones. The whole process is rather frustrating, but eventually, you will end up with the channels you want, containing the talk groups that you are interested in. Although there are thousands of talk groups, only a few are busy.

Using the radio. There is another chapter for that, but basically... Turn on the radio, use the **Menu > Zone > Zone List > Up or Down > Confirm** to select the zone that you want. Each zone includes the channels relating to a repeater or hotspot. Turn the channel knob to select the talk group. Listen to check that the frequency is clear and make a call. Most people do not use CQ on DMR. Say your callsign, "listening for any station on talk group xxx."

PROGRAMMING THE TYT RADIO FOR FM

Programming for FM simplex and repeaters has been covered already, starting on page 86. It is much easier than programming DMR channels. For a start, you don't need to worry about talk groups, time slots, colour codes, or hotspots. You only need one FM channel for each FM repeater. It still has to be included in a zone.

I established a zone for all FM repeaters in New Zealand, and a second zone for our 'National System' of linked FM repeaters. But I am more likely to use our DMR system of 14 ZL-TRBO repeaters.

I placed the standard FM simplex frequencies in the same zone as the repeaters. I have not bothered to include the VHF repeaters, as I have another FM handheld already programmed. But of course, you can add them into the same zone, or a different zone, if you want to.

Programming with EditCP

Dale Farnsworth wrote an excellent Open Source program called 'EditCP' for the TYT MD-380, MD-390, MD-2017, MD-UV380, MD-UV390, and MD-2017, the Alinco DJ-MD40, the Retevis RT3, RT3-G, RT3S, and the RT82 radios. It is available at https://www.farnsworth.org/dale/codeplug/editcp/. EditCP does not work with the Radioddity/AnyTone/BTECH radios.

In many ways, EditCP is easier to use than the official TYT CPS software. This is especially true for the TYT MD-380 and MD-390 models because the TYT software for those models is pretty bad. Its best feature is that it can import and export a spreadsheet containing all the data files. EditCP is not compatible with the Radioddity / AnyTone / BTECH radios and will not read their code plug files.

GETTING STARTED WITH EDITCP

The first thing is to use **File New** to start a new code plug, or **File Open** an existing code plug file. Everything looks greyed out, but the menu structure is working.

EditCP will read the TYT .rdt code plug file saved by the MD-UV380 CPS. But to make sure you are getting the most up to date data, it is probably best to download the structure from your radio using **Radio > Read codeplug from radio** and save that as a starting point, using **File > Save As**. From then on you can open the file using **File > Open Recent** and selecting your code plug file. The program takes a few seconds to load the databases, but eventually, the buttons will activate.

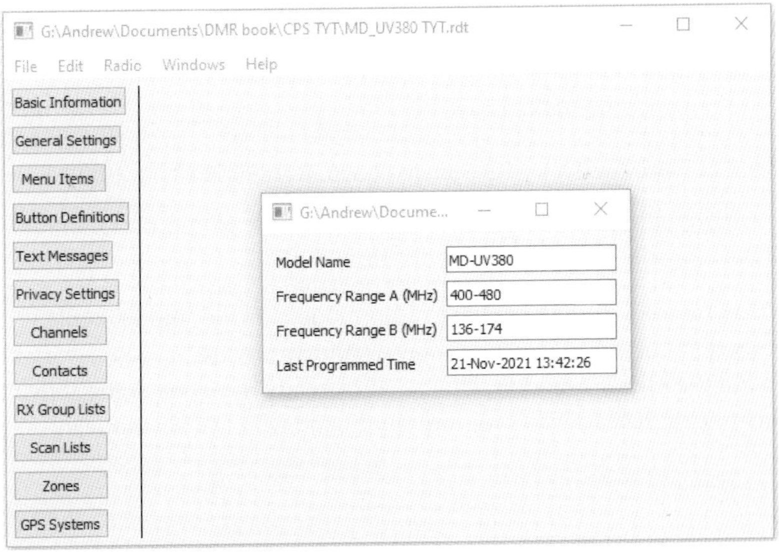

Figure 31: EditCP Basic Information screen

The image above shows the 'Basic Information' for my TYT MD-UV380. It shows the model and frequency ranges, but interestingly it also shows the date and time of the last upload to the radio.

The rest of the buttons open the same sort of screens as the CPS. The Channels, Contacts, RX Group Lists, Scan Lists, and Zones, screen are easier to navigate because the channels, groups, and zones etc. are listed at the left of the screen. All in all, the interface is much smarter and easier to use than the TYT CPS.

CSV contact list (MD-UV380 and MD-UV390)

One thing that EditCP can't do is work with the CSV contact list in the firmware updated MD-UV380 and MD-UV390 models. I think that is because the software was written for the older MD-380 and MD-390 models which don't have that feature. Actually, you can download the contacts data provided it has been placed in the EditCP cache directory as 'usersDB.bin'. But I don't know what format that is, or how to get a .csv file into that format. Maybe it is a feature that is still being developed.

C:\users*name*\AppData\Local\codeplug\Codeplug Editor\cache\usersDB.bin

Finding Bugs in your code plug

When you upload the code plug from the radio or your PC, the program will identify any bugs in your code plug that it finds. This is a very useful feature that can save you puzzling over why something on your radio is not working.

Go to Edit > Show Invalid Fields to show the problems. After you fix them in the various windows, you can go back to Edit > Show Invalid Fields and click the Rescan for invalid values button to check that you have all the bugs ironed out. I had a problem with the one-touch buttons where I referenced a talk group that I had renamed, and I had a digital simplex channel without a Contact Name. Again, because I had renamed the talk group.

Alternative firmware for TYT MD-380, MD-390, Retevis RT3, and RT8.

EditCP can write either the original 'factory' firmware to the radio, or the 'md380 tools' firmware, or the 'KD4Z md380 tools' firmware.

I have no experience with any of these radio models or the 'md380 tools' firmware, but the intention seems to be the same as the firmware update available from Radioddity for the MD-UV models. Unless you already have one of the above models, I would buy one of the easier to program dual-band MD-UV380 models.

CONVERTING YOUR CODE PLUG TO A NEW RADIO TYPE

This is an especially useful feature of EditCP. If you buy another radio model, or you want to give a code plug from your radio to another ham who has a different radio, EditCP may be able to help. FANTASTIC! It is very easy.

Read the code plug from your radio or a saved .rdt file.

Select File > Convert codeplug to new radio type. Use the dropdown list to select the new radio type that you want and its frequency range. For example, the new radio might be a Retevis RT3G on the 400-480 MHz band.

Click OK and the program will immediately create an additional main window with the new code plug. Save it to the PC using File > Save.

Make sure to change the DMR ID if the new code plug is for someone else. Make any other changes while you have the program open. Save it back to the PC using File > Save.

Plug the correct programming cable into the new radio. Yes, it is likely to be a different cable, even if it looks the same. Write the code plug to the new radio. Radio > Write codeplug to radio.

USING THE EDITCP EXCEL FILE

Downloading the file is easy, just use File > Export > Export to Spreadsheet. But for some reason, on my computer, the data is hidden when you open the file in Excel. Very weird indeed!

Open one of the spreadsheet tabs, for example 'Channels.' You will be presented with a blank spreadsheet. Closer inspection reveals that the top row number is not a 1. The rows are hiding from you. The rows are not actually hidden, the row height has been reduced to zero. Why? "I dunno." Click the small triangle to the left of the 'column A' header or click in any cell and use CTRL-A to select all cells. Right-click any row and select Row Height. Enter 14.25 into the text box and click OK. This should restore the missing rows.

Although you get all the data files, it is only worth using the Excel file on the big lists. Note that if you rearrange the order of the 'Contacts' (talk group) list it will break the associations with the 'Contact Name' field in the channels. This is because the TYT radios reference the position of the talk group in the list rather than the talk group number itself. Very bad programming, but it can't be helped.

You can rearrange contacts, zones, and channels, then save the spreadsheet and load it back into EditCP, and from there back into your radio. Note that this duplicates the old code plug. You will get a new window when you import the spreadsheet. File > Import > Import Spreadsheet File.

Radio programming options

There are many options available on your radio, you will have to discover some of them yourself. This chapter covers some of the more useful functions. Some radios do not support all of the listed functions.

- Personal greeting

- Promiscuous mode (not as sexy as you think)

- Hotkey to disconnect

- Default start-up channels

- Using two zones

- FM broadcast channels

- Setting the date and time

- GPS and APRS

PERSONAL GREETING

It is nice when the radio greets you with a cheery "hello" as it boots up. You can customise the bootup greeting on most DMR radios. With the Radioddity, I used HELLO ZL3DW, but with the TYT radio, I decided to keep the rather nice TYT logo.

On the Radioddity/AnyTone CPS, the greeting is set on the **Optional Setting** tab. Change **Power-on Interface** to **Custom Char** and enter your two-line message in the Power-On Display boxes. Or you can select **Default Interface** for the standard icon or **Custom Picture**. To add a custom picture, set the image to be used in **Tool > Boot Image**. The image should be a square .bmp or .jpg file. The software will compress any image that is not square. You have to **write** the image to the radio as well as selecting the image in the Optional Setting tab. Then you have to do the usual **write to the radio**. With the TYT CPS, you can choose from the TYT logo **Picture** or a two-line **Char String** text message. The setting is in the **Intro Screen** section of the **General Setting** tab.

DIGITAL MONITOR (PROMISCUOUS) MODE

The promiscuous mode was a feature of early modes such as the MD-380 and MD-390. On the new models, it is called **Monitor** or **Digital Monitor**. It is usually associated with one of the buttons above or below the PTT button on the radio. It allows you to hear all of the traffic on the current channel's time slot. The AnyTone/Radioddity also has a dual monitor, so you can hear everything on the repeater or hotspot

USING A HOT KEY TO DISCONNECT A TALK GROUP

You can use the Hot Key function to allocate the talk group disconnect function to one of the three programmable function keys or a 'function key.' I changed the short press function of the PF1 key to 'disconnect' because it is immediately below the PTT switch.

TIP: You only need to disconnect a talk group if it is a dynamic talk group and it was you who established the link, not a static talk group that is permanently allocated to the repeater or hotspot. You can leave your personal hotspot associated with any talk group(s) you like.

For the Radioddity GD-AT10G / AnyTone AT-D878UV radios.

1. Select **Common Setting** > **Hot Key** and click on the **Hot Key** tab

2. On the Hot Key 1 line, double click 'Menu' in the Mode column and use the dropdown list to change the entry to **Call**.

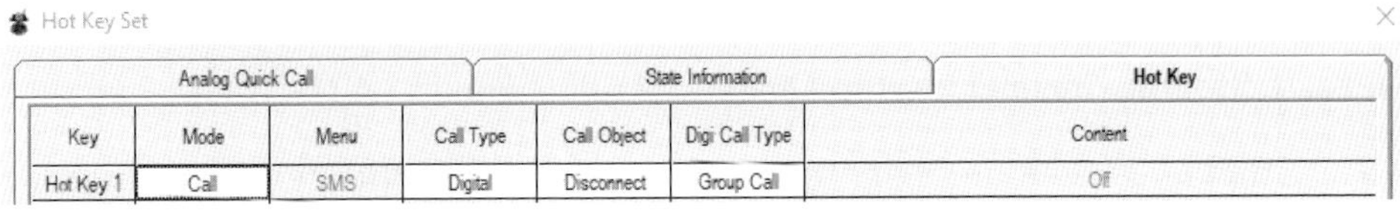

Hot Key Set							✕
Analog Quick Call				State Information		Hot Key	
Key	Mode	Menu	Call Type	Call Object	Digi Call Type	Content	
Hot Key 1	Call	SMS	Digital	Disconnect	Group Call	Off	

Figure 32: Setting up a Hot Key

3. Double click 'Off' in the Call Object column and select your 'disconnect talk group.' Mine is just called 'Disconnect.'

4. Click **OK** to save your changes.

    *TIP: The function keys are activated by pressing and holding the **Menu** (-) button until the display says, 'Next Please Press Dial Key,' then press the required button (0-9, # or *).*

5. You will need to extend the time that the function is available. Select **Common Setting** > **Optional Setting** and the **Digital Func** tab. Change the Group Call hold **Time** to around 25 seconds. This will give you time to press PF1 to change to the 'disconnect talk group' and then activate the PTT to disconnect the talk group.

For the TYT radios.

The two side buttons can be similarly configured on the **Buttons Definitions** tab on the TYT CPS. You can even set how long you have to press a side button to make it a 'long press.' You can also set 'one-touch access' and assign talk groups to 'number key quick contact access.' But there is nothing in the "manual" that says how you use the one-touch buttons.

DEFAULT START-UP CHANNELS

The TYT radios will start on the zone and channels that were active when you turned off the radio. If you load a new code plug from the CPS, the radio will start on the first zone in the zone list, with the channels that are at the top of the channel list.

You can configure the Radioddity/AnyTone radios to start up with the same channel or channels on the display every time, or with the last used channel. I set my radio so that it starts with my Brandmeister monitor channel (TS1) on the top and the DMR+ monitor channel (TS2) on the bottom of the display. I usually only display one channel set to my Brandmeister monitor channel. A quick tap of the P1 button changes over to my DMR+ monitor channel. A long press of the P1 button activates both at the same time.

On the Radioddity/AnyTone CPS, you can set what you want on the **Common Setting > Optional Setting > Power On** tab.

If **Default Startup Channel** is set to **Off**, the last used channel will be used.

If **Default Startup Channel** is set to **On**, the default channels will be used.

- **Startup Zone A** and **Startup Channel A** sets the zone and channel for the primary (upper) display.

- **Startup Zone B** and **Startup Channel B** sets the zone and channel for the secondary (lower) display.

USING TWO ZONES

It is possible to set the upper channel to a different zone from the lower channel. On the TYT radios, you will get a choice of the 'channel A' channels for zone 1 on the top and a choice of the 'channel B' channels for zone 2 on the lower display. On the Radioddity/AnyTone radios, there is no A and B list, so you will get the channels you set for each zone.

FM BROADCAST CHANNELS

The Radioddity/AnyTone radios can receive FM broadcast stations. The TYT radios can't. I thought this was interesting given that the Radioddity is a UHF radio. Well above the FM broadcast band. The FM receiver is quite difficult to turn on. It is well hidden under several menu layers. Also, the minimum volume is very high. I guess it could be handy if you are hiking, but I can't see much value in the FM radio.

To turn on FM, **Menu > Settings > Radio Set > Other Func >FM Radio > Radio On.** You can shortcut this by setting the FM radio to one of the function keys. The channel frequencies are set in the CPS **Common Setting > FM**. They cannot be set using the keys on the radio. The channel knob changes channels.

The **FM Radio Moni** setting allows you to receive and transmit on the currently selected DMR channel as well as listening to the FM broadcast channel. When it is off you cannot use the DMR functions at all.

SETTING THE DATE AND TIME

You cannot set the date and time from the CPS. It has to be done on the radio.

TYT Date and time

I do not have a GPS equipped TYT radio, so I don't know if the GPS sets the clock on the models that have GPS. I expect that it does because there is a UTC 'Time Zone' offset setting in the CPS General Setting tab. Unfortunately, it does not have a UTC+13 setting so it cannot be set for New Zealand Summer Time. So maybe it is just as well I don't have the GPS model.

Set the date and time using **Menu > Utilities > Radio Settings > Clock > Time > Confirm**. Then enter the time, using the up and down keys to move across the digits. Press (Menu) **Confirm** to save the **Time**.

Hit the red **Back** button so you can select **Date > Confirm**. Then enter the date, using the up and down keys to move across the digits. Press **Confirm** to save the **Date**.

Hit the red **Back** button so you can select **TimeZone**. Use the up and down keys to select the correct UTC offset and (Menu) **Confirm**. The Time Zone is only used for date stamping messages and recordings.

Radioddity/AnyTone date and time

On the radio select, **Menu > settings > Radio Set > Other Func > Time Zone**. Use the up ▲ arrow or down ▼ arrow the select the Time Zone (UTC offset). Press the Menu button to **Confirm**. Unlike the TYT radio, the Radioddity/AnyTone radios do allow for a UTC+13 time zone.

Set **Menu > Settings > Radio Set > Other Func > Time Display** to **on** if you want the radio to display the date and time.

Menu > Settings > Radio Set > Other Func > Date Time

↳ **Time Set** Use the ▲ arrow or down ▼ arrow change the date and time numbers. The P1 button moves you to the next number. Only do this if you are not using the GPS to discipline the clock. If the GPS is on and locked (red splodge), use GPS Check to set the clock.

↳ **GPS Check**. You have to set the clock to accept the GPS discipline.

↳ **Dis Formart (sic)**. This lets you choose between the normal date format and the one the Americans use. Select dd/mm/yyyy or mm/dd/yyyy.

Radioddity/AnyTone using GPS to set the date and time

The Radioddity/AnyTone radios have GPS as standard, but it has to be enabled in the CPS before it will correct the clock. First, you have to turn the APRS and GPS features on. Select Tool > Options and check GPS and APRS. Only check Analog APRS RX if you plan to use FM APRS. If you haven't already, select Common Setting > Optional Setting > GPS/Ranging > GPS > On.

On the radio, after the red splodge in the top centre of the display indicates that GPS is working, you have to set the clock to accept the GPS discipline. Menu > Settings > Radio Set > Other Func > Date Time > GPS Check.

RADIODDITY/ANYTONE GPS

The Radioddity/AnyTone radios have GPS as standard, but it has to be enabled in the CPS (or on the radio) before it can be used and before it will correct the clock. Select Common Setting > Optional Setting > GPS/Ranging > GPS > On.

On the radio, a blue splodge at the top centre of the screen indicates that GPS is on, but a GPS lock has not been established. A red splodge indicates that a GPS lock has been achieved. After the red splodge appears you can select Menu > GPS > GPS Info for a display showing your location and the accurate date and time.

Do not turn on the next item 'Get GPS Positioning.' It sends an SMS message to the radio that you are calling which then responds with its position. This is not supported on most DMR networks, and it will only work if the other radio has the same function.

The GPS > Start Test option is not mentioned in either the Radioddity manual or the AnyTone manual. If you turn the test function on, the radio will do a GPS test when you turn on the radio. It displays information received from the GPS satellites. You will see a sudden burst of positioning data after the radio has received data from two or more satellites. After about 30 seconds the display reverts to the normal start-up screen. This test has no value, so leave it turned off.

APRS

The Radioddity/AnyTone radios have APRS and GPS as standard. Note that many DMR networks do not support APRS because the APRS beacons cause hundreds of repeaters to operate all over the world. The networks that do support APRS only allow its use on specific talk groups. On a channel set up for that talk group, you can check the APRS RX check box in the CPS channel setup screen. You will be able to see the location details of people sending APRS packets over the APRS talk group. See the APRS chapter for the details.

Radio Setup TYT

I don't have room to write a complete manual for all the different TYT radios, but I will include the basic operation of the TYT MD-UV380 in this chapter. Other TYT radios should be similar. The TYT-UV390 is identical. The TYT MD-UV390 is a waterproof version of the MD-UV380 and the MD-390 is a waterproof version of the MD-380. The MD-380 is not the same radio as the MD-UV380. It is a single band radio that lacks many of the features included in the UV models. The same is true of the MD-390 and the MD-UV390. The CPS for the MD-380 and MD-390 is horrible. There is virtually no way to upload files from your PC. The CPS for the UV models is OK, but not as good as the CPS for the AnyTone/Radioddity radios. It has less flexibility, fewer options for using spreadsheets, and the uploads and downloads to the radio are slower.

TIP: In my opinion, the TYT MD-UV380 and MD-UV390 radios are "just not as good," as the AnyTone/Radioddity/BTECH radios. They don't have as many features, and they are much more difficult to configure. I am much happier with the Radioddity GD-AT10G.

COMPARISON WITH THE RADIODDITY GD-AT10G

- The MD-UV380 is slightly smaller and fits nicely into your hand. This is due to the bigger battery on the Radioddity radio.

- The supplied equipment was the same – but make sure that the package includes the programming cable and the charger. Not all of them do.

- The TYT radios are dual-band, and dual-mode DMR and FM. The GD-AT10G is a single-band radio but the AnyTone AT-D878UV is a dual-band radio

- 5 watts or 1 watt power. The GD-AT10G also has a 10-watt setting

- 2000 mAh battery, the Radioddity battery is 3100 mAh

- The TYT weighs 278 grams, the Radioddity weighs 322 grams (16% heavier)

- GPS is not available on all models and is not worth buying anyway

- 3000 channels, 64 channels per zone, (only 16 on an MD-380 / MD-390).

- 10,000 contacts including talk groups. Upgradeable to 120,000 contacts with a firmware upgrade. You lose the recording function to make space for the bigger contact list. This upgrade is recommended as you get much better caller information. The GD-AT10G can store 200,000 contacts.

- The TYT radios always show two channels on the display. This is optional on the AnyTone, Radioddity, Alinco, and BTECH radios. However, you can mute either channel.

TYT FIRMWARE UPGRADE

The standard factory firmware supplied on my TYT radio only allows for 10,000 contacts, and that includes the talk groups. Also, the Private ID data that is held in the contact list only includes the person's name and DMR ID. This means that their callsign is not displayed, which is disappointing. There are about 206,000 registered DMR users so my radio could only identify around 5% of DMR users. The rest come up as 'Unknown ID.'

You can apply a firmware upgrade which adds the ability to upload a .csv file into the CPS program and from there into the radio. When this is uploaded and activated, the radio will display the calling person's, Radio ID,   Callsign,   Name,   Nick Name, City, State, and Country. The 'CSV' list can hold 120,000 contacts, so now the radio can identify 58% of possible callers.

Rules

The CSV list only holds information for displaying the details of heard stations. It is not a talk group or contact list. If you are using a CSV list, you should create a contact list that only holds the talk groups that you want to use and any private call IDs that you want to talk to. My contact list contains 1 Private call ID and 28 talk group IDs.

- The CSV contact list should not include talk groups

- The standard contact list should only include talk groups and perhaps one or two private call IDs

WARNING turning the radio off during a firmware upgrade, or the upgrade software crashing, can turn your radio into a brick. I had no problem doing the firmware upgrade, but if you choose to do so, it is entirely at your own risk.

That said, I do recommend putting on your special "brave shoes" and doing the following firmware upgrade.

Firmware versions

You can find the current firmware revision by pressing **Menu > Utilities > Radio Info > Versions**. My current version is V018.011.

You cannot upgrade a non-GPS model to a GPS model by changing the firmware. Non GPS models don't have a GPS receiver in them.

TIP: No firmware upgrade is necessary on the Radioddity/AnyTone/BTECH radios. They already hold 200,000 contacts and they have a separate talk group list.

Firmware upgrade process

Visit https://www.radioddity.com/pages/tyt-download and click on the picture of the MD-UV380. Don't worry if you have a UV390 or about GPS. In the popup window, click the link to download the MD-UV380 Firmware V18.11 (or later revision). This will download a zip file containing two CPS program installers, four firmware .bin files, a Word instruction document and an application called 'FirmwareDownloadV3.04_EN.exe.' Extract the zip to a directory on your computer.

Firmware versions for MDUV380 and MDUV390		
First letter	Installed	Firmware file
D	D018.011	Factory - No GPS 10,000 contacts + recording
V	V018.011	120,000 contacts, no recording
P	P018.011	Factory - GPS 10,000 contacts + recording
S	S018.011	120,000 contacts, no recording

- If your radio does not have GPS, upgrade it from version D to version V

- If your radio has GPS, upgrade it from version P to version S

Install the 'UpgradeDownload' program by double-clicking the Firmware DownloadV3.04_EN program file. You will get the same dire warnings from Windows 10 as you did when you installed the CPS program.

To run the program after it has been installed, use the search icon on the Windows toolbar to find 'UpgradeDownload' or navigate to the install directory and fire it up from there. I right-clicked the .exe and created a shortcut which I dragged into my DMR directory. I also moved the .bin file there.

Click **Open file upgrade** and select the new firmware file.

Version V is for radios without GPS and S is for radios with GPS. The other two are in case you want to go back.

Now we have to prepare the radio. Turn the radio off. Connect the programming cable between the radio and your computer as per usual. On the radio hold down both the **PTT and the button above the PTT** and turn on the radio. The LED should alternate red and green. Stop pressing the buttons, do nothing else on the radio, and put it down on the desk. The flashing red, green indicates that the radio is in bootloader mode.

Figure 33: UpgradeDownload firmware updater

Click the 'Download file of upgrade' button on the 'UpgradeDownload' firmware program. A green indicator bar on the software will show that the download is happening. Remember the warning above. The indicator will increase until it reaches about 50% then it will stop for about 60 seconds (maybe more). Don't panic - this is normal. Finally, the radio will reboot and should return to the normal display. Breathe!

If the radio does not restart, try turning it off and even removing the battery for a few seconds. Then power up again and hope for the best. If that fails, try to re-flash the new firmware. If that fails, try to re-flash the original firmware. Are you sure about the GPS version? If that fails, panic. I experienced no problems, everything went perfectly.

USING THE CSV LIST

The new firmware adds a new item to the radio menu. You need to turn on the CSV contact list option on the radio, so that will display caller data.

Menu > Utilities > Radio Settings > ContactsCSV > Turn On > Confirm

The CSV list is only used for displaying caller data, it does not affect the operation of the radio. But it does mean that there is no need to put Private call IDs in the contact list unless you want to make a private call to that person. Loading the CSV list into the CSP program is a bit odd. It feels like a 'kludge' added on the side.

1. Create a CSV contact list from data downloaded from RadioID.net, or AmateurRadio.digital, or build one yourself. The columns should be headed, Radio ID, Callsign, Name, Nick Name, City, State, and Country. The 'CSV' list can hold up to 120,000 contacts. Do not put talk groups into the CSV list.

2. Save and close the spreadsheet as UV380_CSV contacts.csv or something similar.

3. Open the TYT CPS. Under **Program** at the top of the page, select **Write Contacts**. Yes, I know that makes no sense.

4. Click **Import** and select your UV380_CSV contacts.csv file.

5. Nothing much will happen, but the file will load. There is no way to view or edit the file in the CPS. When the text on the buttons goes black, the file has been uploaded.

Figure 34: Write Contacts entry screen

6. Connect the programming cable in the usual way, then click the **Write** button. A green bar indicates the download process. The download can take several minutes. Go and get a coffee.

7. Nothing will look different on the radio or the CPS software. But when a caller that is in the database comes in via the repeater or hotspot you should see a full description. Note that only around half the possible users are there.

You can view the CSV list on the radio, but good luck scrolling down to entry number 100,000. **Contacts > View Contacts CSV > Contacts CSV List.**

Or if you want to look up a particular ID, enter **Contacts > View Contacts CSV > Enter Number > 1234567 > Confirm.** If the contact number is in the list, it will be displayed. If it is not in the list, the error message says '**ContactsCSV Empty**' which is initially a bit disconcerting. The first two times it happened; I unnecessarily reloaded the list!

TIP: I had a problem with the list because all the ZL entries were greater than number 120,000 in my source database. Indexing the .csv file and even moving the ZLs to the top had no effect because the CPS program re-indexes the list in order of DMR ID numbers. Of course, all the VKs and ZLs start with a 5 so they were too far down the list and were not being identified.

I had to trim the list by deleting contacts from countries above 505 that I probably won't work, and several thousand American stations (sorry).

DIGITAL SIMPLEX

Setting up simplex frequencies for FM has already been covered. There are two ways that you can make a digital simplex radio-to-radio call. You can make a Group Call on TG 99 'Simplex' to another radio also set for TG 99 'Simplex.' Make a channel on your nominated simplex frequency and TS1 for the TG 99 talk group. In New Zealand, the nominated simplex frequency for digital voice modes is 432.750 MHz.

This type of call will reach any DMR radio that is within range and on the same channel and talk group. The radio ignores the time slot setting, but the colour code must match.

The other way is to make a Private ID call to a radio which uses that ID. The other radio needs to be programmed with a channel to make a Private ID call to your DMR ID. I tried this out with both radios on my DMR ID and it worked fine. You can nominate a call alert to ping when you get called. And you can send out an enquiry to see if the other radio responds. This type of call will only reach the nominated DMR radio, assuming that it is on the same channel and within range.

DISPLAY ICONS

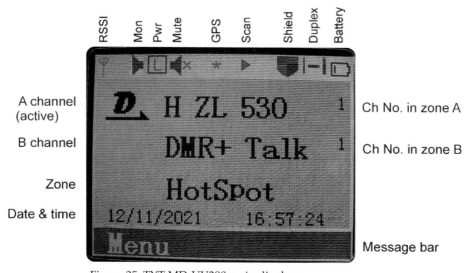

Figure 35: TYT MD-UV380 main display

- Shield icon blue/red = un-encrypted, blue/yellow = encrypted.
- Duplex icon: |-| means duplex, ➡ means Simplex.

MENU TREE FOR THE TYT MD-UV380

Contacts			
↳	Contacts	→	Select a contact
↳	New Contact	→	Enter a new contact
↳	Manual Dial		
	↳	Group Call	Enter TG number
	↳	Private Call	Enter Private ID No.
↳	View Contacts CSV		
	↳	Contacts CSV List	Displays callsigns
	↳	Enter Number	Enter Private ID No.

Scan			
↳	Turn on (off)	→	Turns the scan on or off
↳	View List	→	Shows scan list channels
↳	Add CH	→	Add channel to the scan list
↳	Scan List	→	Select a Scan List
↳	New Scan List	→	Add a New Scan List
Zone			
↳	Zone List	→	Select a zone
↳	New Zone	→	Create a new zone

Messages			
↳	Inbox	→	View list
↳	Write		
	↳	Send	Edit and send
	↳	Save	Edit and save as draft
		Clear	Edit and save as draft
↳	Quick Text		
	↳	Send	Edit and send
	↳	Save	Edit and save as draft
	↳	Clear	Edit and save as draft
↳	Sent Items		View list
↳	Drafts		
	↳	Drafts	Edit and send
	↳	Delete All	Delete drafts

Call Log			
↳	Missed		
		↳	Missed List	List missed calls
		↳	Delete All	Delete list
↳	Answered		
		↳	Answered List	List answered calls
		↳	Delete All	Delete list
↳	Outgoing		
		↳	Outgoing List	List outgoing calls
		↳	Delete All	Delete list

Utilities			
↳	Radio Settings		
1		↳	Talk Around	Reverse repeater
2		↳	Tones/Alerts	Set tones & alerts
3		↳	Power	Set RF power
4		↳	Backlight	Set backlight time
5		↳	Squelch	Set squelch level
6		↳	Intro Screen	Picture or text
7		↳	Keypad Lock	Lock keyboard
8		↳	LED Indicator	Turn off LED
9		↳	Password Lock	Set a password
10		↳	Clock	Set date & time
11		↳	Mode	Channel or memory
12		↳	Voice Announcement	On or off
13		↳	Contacts CSV	Set to on
14		↳	Private Call Match	Set to on
15		↳	Group Call Match	Set to on
16		↳	Menu Hang Time	Menu delay
17		↳	TX Mode	Last Ch

Utilities			
↳	Radio Info		
1	↳	My Number	Call and DMR ID
2	↳	Versions	Firmware versions
↳	Program Radio		
1	↳	RX Frequency	Edit RX Freq
2	↳	TX Frequency	Edit TX Freq
3	↳	Channel Name	Edit CH name
4	↳	Time Out Timer	Set TOT
5	↳	CTCSS/DCS	Set CTCSS tone number
6	↳	Add CH	Add a new Channel
7	↳	Color Code	Set CC
8	↳	Time Slot	Set TS
9	↳	VOX	VOX trigger 1-10
10	↳	TX Contact	Select a talk group
11	↳	Group List	Select or new
12	↳	Mic Level	Set mic level 1-6

MUTING A CHANNEL

The TYT radio always displays two channels (A & B). They can be on different time slots, zones, talk groups, repeaters, or hotspots.

The receiver scans each channel until it receives a signal. You can see this with the green speaker icon (◀) alternating between the channels on the left side of the channel names. This can be a major problem if a call comes in on one of the channels when you want to listen to the other. You can get frozen out. If the upper channel is active, you may be able to change the channel switch to stop hearing the unwanted call.

The D icon indicates which channel you will transmit on if you press the PTT button. Press in the red 'back' button when the D icon is on the top channel and it will mute the lower channel, (red ◀x icon).

Press in the red 'back' button when the D icon is on the lower channel and it will mute the upper channel, (red ◀x icon).

TIP: It is not a bad idea to mute the second channel while you are talking to someone so that you don't get frozen out of your QSO. The alternative is to set the alternate channel, A or B, to a DMR channel that has very little activity.

Radio Setup Radioddity/AnyTone

I don't have room to write a complete manual for all the different Radioddity, AnyTone and BTECH radios, but I will include the basic operation of the Radioddity GD-AT10G in this chapter. The other radios based on the D878U framework use a very similar CPS and operation. I believe that the AnyTone AT-D878UV is a re-badged dual-band version of the GD-AT10G it looks the same. Although the Alinco DJ-MD5 looks different, it uses a very similar CPS. The BTECH-6X2 has some firmware differences, but it looks exactly the same as the Radioddity GD-AT10G.

I like the Radioddity radio much more than the TYT MD-UV380. It works better, has better features, has faster upload and download speeds, and the CPS software is significantly better. GPS is supplied as standard. It keeps the clock accurate, and the radio does support APRS. The TYT radio needs a firmware upgrade just to get close to the same performance.

COMPARISON WITH THE TYT MD-UV380

- The Radioddity GD-AT10G is slightly fatter because of the bigger battery

- The supplied equipment was the same – but make sure that the package includes the programming cable and the charger. Not all of them do.

- The TYT MD-UV380 and MD390 are dual-band radios, The Radioddity GD-AT10G is a single band UHF radio, which is fine for DMR but may be a problem if you want to access FM repeaters. The AnyTone AT-D878UV is a dual-band radio, but it is more expensive.

- All DMR radios are dual-mode DMR and FM

- 10 watts, 5 watts, or 1 watt power

- The TYT weighs 278 grams, the Radioddity weighs 322 grams (16% heavier)

- Larger 3100 mAh battery

- GPS is standard and the radio supports APRS

- 4000 channels, 200,000 contacts. (No need for a firmware upgrade). The TYT MD-UV380 and MD390 supports 10,000 contacts, firmware upgradeable to 120,000 contacts.

- The last heard function is great. It displays the talk group, name, and callsign of the last call the radio heard. It is good if you are using a monitor channel because it tells you which channel to switch to and the callsign to call.

- The dual slot monitor lets you listen to both time slots.

MENU TREE FOR THE RADIODDITY GD-AT10G

The Radioddity radio has a much more complex menu tree than the TYT MD-UV380, with many more features. Most of the time it is better to use the CPS, but occasionally you may want to make adjustments without a PC connection.

Talk Group			
↳	TG List		Select a contact
↳	New Contact		Enter a new contact
	↳	Input ID	
	↳	Input Name	
	↳	Input Address	
	↳	Input Callsign	
	↳	Input State	
	↳	Input Country	
	↳	Input Remarks	
	↳	Ring Set	
	↳	Save	Save
↳	Manual Dial		
	↳	Private Call	Enter Private ID No.
	↳	Group Call	Enter talk group No.
↳	Talker Alias		
	↳	Alias TX Set	Not widely used
	↳	Alias RX Display	Not widely used

SMS			
↳	New Msg	→	Send or save
↳	InBox		
	↳	InBox List	List inbox
	↳	Delete All	Delete all messages
↳	OutBox		
	↳	OutBox List	List or edit drafts
	↳	Delete All	Delete all messages
↳	Quick Text	→	Select from list
↳	Draft		
	↳	Draft List	List or edit drafts
	↳	Delete All	Delete all messages

Call Log			
↳	Last Call		
	↳	Last Call List	Lists calls, save to contacts
	↳	Delete All	Delete list
↳	Sent Items		
	↳	Sent List	Lists calls, save to contacts
	↳	Delete All	Delete list
↳	Answered		
	↳	Answered List	Lists calls, save to contacts
	↳	Delete All	Delete list
↳	Missed		
	↳	Missed List	Lists calls, save to contacts
	↳	Delete All	Delete list

Zone	→	Lists zones	Select a zone

Scan			
↳	Scan On/Off	→	Start or stop a scan
↳	Cdt Scan	→	FM CTCSS SQ
↳	Scan List	→	Add or set Priority

Roaming			
↳	One Time Roam	→	looks for DMR repeaters
↳	Roaming Zone	→	Add or edit a roaming zone
↳	Auto Roaming		
	↳	On / Off	Turn on auto-roaming
	↳	Fixed Time Set	Roam after x mins if no repeater is found
	↳	Start Roaming	At a fixed time or out of range
↳	Repeater Check		
		On / Off	Turn repeater check on
		Intervals Set	Check repeater every x secs
↳	Out Range Note		
		Note Kind	Beep or 'Sound'
		Note Times	How many beeps?

↳	Effect Wait	→	Unknown
Settings			
↳	Radio Set		
1	↳	Voice functions	15 options
2	↳	Display functions	14 options
3	↳	Key functions	11 options
4	↳	Other functions	29 options
↳	Channel Set		[FM options are different]
1	↳	New Channel	Create a channel
2	↳	Delete Channel	Delete a channel
↳	Channel Set		
3	↳	Channel Type	Analog or Digital
4	↳	TX Power	Low Med High Turbo
5	↳	Offset	Usually 0.00 Hz
6	↳	Bandwidth	Narrow for DMR
7	↳	RX Frequency	Set (P2 to exit)
8	↳	TX Frequency	Set (P2 to exit)
9	↳	Talk Around	On or Off
10	↳	Name	Edit (P2 to exit)
11	↳	TX Allow	Channel Free
12	↳	TX Prohibit	Off unless RX only
13	↳	Radio ID	View or add
14	↳	Color Code	Set for the repeater
15	↳	Time Slot	Set for the talk group
16	↳	Digi Encrypt	Off
17	↳	Encryp Type	Normal
18	↳	RX Group List	choose a list
19	↳	Work Alone	Off
20	↳	CH Ranging	Off
21	↳	APRS Receive	Optional
22	↳	SMS Forbid	Off
23	↳	DataAck Forbid	Off
24	↳	DMR Mode	Repeater or Simplex only
↳	Device Info	→	Displays radio ID and name

Record			
↳	Record Switch	→	Turn recording on
↳	Record List	→	Play, send, or loop playback
↳	Record Delete	→	Delete all recordings

Digi Moni			Digital Monitor
↳	Digi Moni Switch	→	Off Single Dual Slot, also on the PF2 button
↳	Digi Moni CC	→	Same or Any CC (usually any)
↳	Digi Moni ID	→	Same or Any ID (usually any)
↳	Slot Hold	→	Dual Slot mode. Radio waits rather than transferring to the other Time Slot

APRS			
↳	Upload type		
	↳	Analog APRS	For an FM APRS repeater
	↳	Digital APRS	For a DMR talk group
↳	Analog APRS		
	↳	PTT upload	Off, PTT, or PTT release
	↳	Upload Power	Set RF power for beacon
	↳	Upload WN	Wide for FM
	↳	Upload Freq	APRS simplex frequency P2 Exit
	↳	Signal Path	APRS path Wide1-1 (P2 exit)
	↳	Upload Text	Text to send with APRS data
↳	Digital APRS		
	↳	PTT upload	Off, or PTT release
	↳	Report Channel	Select. (Set in CPS 1-8)
	↳	Upload Slot	1,2, or per the channel
	↳	Edit ID	Edit TG no and private/group
↳	Digi APRS Info	List	View received APRS data
↳	Intervals Set	→	Set beacon interval
↳	Upload beacon	→	Fixed or GPS (use GPS)

APRS

You can pay extra to get a TYT MD-UV380 or MD-UV390 with GPS. Some of the adverts are misleading, hinting that the radio has a GPS receiver when the particular radio being advertised does not. The firmware is different on the GPS models, but there is no easy way to tell if one has a GPS receiver. I'm not sure if GPS will keep the clock accurate. It is not mentioned in the manual, but I expect that it does. The GPS takes a long time to lock and is not very sensitive. APRS is not supported on TYT radios, but if you use the Brandmeister network you can send GPS beacons to the APRS.fi website over TG 310999 (USA) and 234999 (UK). The Brandmeister server adds the APRS text string and icon.

The Radioddity/AnyTone radios have APRS and GPS as standard. Note that many DMR networks do not support APRS because the APRS beacons cause hundreds of repeaters to operate all over the world. The networks that do support APRS only allow its use on specific talk groups. UK Phoenix supports APRS on TG 5057, DV Scotland uses TG 9057, and Brandmeister supports it on TG 234999 for the UK and 310999 for the USA. On a channel set up for APRS, it is OK to check the APRS RX check box in the CPS channel setup screen. You will be able to see the location details of people sending APRS packets over the APRS talk group. *TIP: this seems to work on 310999 but not on 234999.* APRS beacons are shown on the Internet at APRS.fi.

FM APRS

The radio supports FM APRS and DMR APRS. FM APRS can send beacons to an APRS digipeater. They are Simplex channels because there is no need to re-transmit your beacons over the local area. I can't test this mode because our APRS channel is on 144.575 and my Radioddity is a UHF only model.

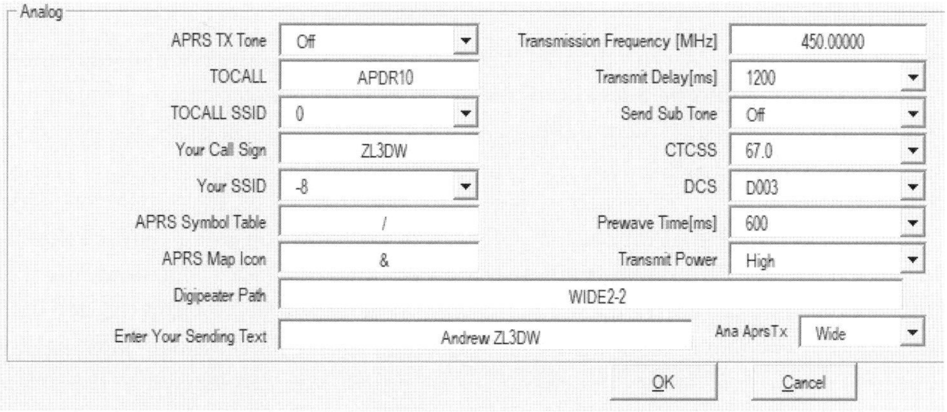

Analog				
APRS TX Tone	Off	Transmission Frequency [MHz]	450.00000	
TOCALL	APDR10	Transmit Delay[ms]	1200	
TOCALL SSID	0	Send Sub Tone	Off	
Your Call Sign	ZL3DW	CTCSS	67.0	
Your SSID	-8	DCS	D003	
APRS Symbol Table	/	Prewave Time[ms]	600	
APRS Map Icon	&	Transmit Power	High	
Digipeater Path	WIDE2-2			
Enter Your Sending Text	Andrew ZL3DW	Ana AprsTx	Wide	

OK   Cancel

Figure 36: FM APRS setup

These are typical settings for APRS over an FM channel to be transmitted to an APRS simplex digipeater.

APRS ON THE RADIODDITY/ANYTONE RADIOS

First, you have to turn on the APRS and GPS features in the CPS. Select Tool > Options and check GPS and APRS. Only check Analog APRS RX if you plan to use FM APRS. This will enable the Common Setting > APRS screen in the CPS which contains most of the settings for using APRS on your Radioddity/AnyTone radio.

You can configure up to eight channels for APRS over DMR. APRS channels on different repeaters may have the same name. Note that many repeater owners ban the use of APRS (including the ZL-TRBO network in New Zealand). So it is better to experiment with APRS channels on your hotspot.

No.	Report Channel	Report Slot	APRS TG	Call Type
1	BM ZL 530	Slot1	234999	Private Call
2	APRS UK	Channel Slot	234999	Private Call
3	APRS USA	Channel Slot	310999	Private Call
4	Current Channel	Channel Slot	0	Private Call
5	Current Channel	Channel Slot	0	Private Call
6	Current Channel	Channel Slot	0	Private Call
7	Current Channel	Channel Slot	0	Private Call
8	Current Channel	Channel Slot	0	Private Call

Repeater Activation Delay[ms] 200

Figure 37: Settings for APRS over DMR

In Report Channel, select the channel that you want to transmit on when you send an APRS beacon. It could be a local talk group on a local repeater channel if you are a hiker, and if the repeater owner allows APRS. It could be Local TG 9. Or it could be one of the dedicated APRS talk groups or a worldwide talk group so that overseas callers can see your APRS beacons. See APRS options below. If you leave 'Current Channel' selected, the APRS beacon will be sent anytime you use a channel that has the Digital APRS PTT mode turned on, or according to the APRS beacon setting. I am not keen on using the Current Channel setting, I prefer to use APRS on the dedicated talk group. If you leave Report Slot set to Channel Slot, the APRS will automatically choose the same time slot as the channel is using. So, you can leave it at the default setting.

APRS data is not sent over the talk group you are using. It is transmitted after you release the PTT switch on the talk group set in the APRS TG box. UK Phoenix supports

APRS on TG 5057, DV Scotland uses TG 9057, and Brandmeister supports it on TG 234999 for the UK and 310999 for the USA. There may be an APRS talk group on your local network, but I don't know of any others.

TIP: If you monitor Brandmeister talk group 234999 or 310999 and you have APRS RX turned on in the channel setup, you may see the APRS information when others send an APRS beacon on the talk group. APRS will not send unless the radio has a GPS lock, or 'Fixed Location Beacon' is turned on.

Required for all APRS channels

You need to set up this area for all APRS channels, (FM or DMR). It sets the basic radio parameters for APRS.

Manual TX Interval[s] sets the interval in seconds between sending APRS information by operating the PTT switch. I recommend setting it to at least **90 seconds**. Otherwise, the radio will send APRS data every time you press the PTT.

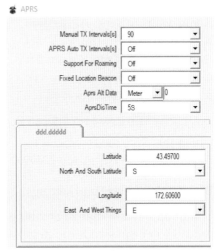

APRS Auto TX Interval[s] sets the interval between automatically sent APRS beacons. I recommend that you turn this **off** unless you specifically want to show a track on APRS.fi. If you are walking set the time to five or ten minutes. If you are driving set the time to one or two minutes. There is no point in sending automatic beacons if you are sitting at home. One PTT triggered beacon is enough. You can watch it come up on APRS.fi.

Figure 38: Settings for all APRS

I left Support for Roaming turned **off**. I guess if you are driving from one repeater coverage area to another then it is worthwhile setting up roaming.

Leave Fixed Location Beacon turned **off** so that the radio uses your GPS location. If you turn it on, the radio will send the coordinates in the ddd.ddddd area. You don't have to fill this in. I did, just to show the format that you would use.

APRS Alt Data sets your altitude in metres or feet. Set your altitude in the box to the right of the dropdown. It appears that unlike D-Star your altitude is not part of the GPS data. It did not come up on my beacons and it is not shown on the radio GPS info screen.

APRS Dis Time sets how long the APRS data is displayed on the radio.

Channel settings

Ah-ha! You thought you were finished! But no. You have to make a change to the channel that you will use for APRS transmissions. As I said above the radio does not transmit APRS on the selected channel talk group. It sends your speech over the talk group as normal and when you release the PTT it sends an APRS packet to the APRS talk group that you selected on the APRS tab on the CPS. If you have created a channel specifically for APRS on TG 234999 or TG 310999, you will just click the PTT to send the APRS beacon.

In the CPS, go to the channel that you selected for sending APRS. You will see that there are four APRS options, that were not there before you enabled APRS in the CPS.

Set APRS report type to **Digital**.

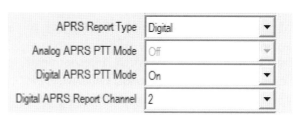

If you want to send APRS data when you release the PTT, set Digital APRS PTT mode to **on**. Set it to off if you only want to use regular beacon transmissions.

Set Digital APRS Report Channel to the **Report Channel** (1-8) that you set on the APRS tab. In this case, the channel is my APRS UK channel, so the Report Channel is No 2, APRS UK APRS talk group 234999.

Brandmeister Self Help settings

You don't have to change anything on the Brandmeister Self Help page. The APRS beacon will be passed through to APRS.fi.

However, you can change the APRS icon, APRS callsign extension, and the 20 characters of text that is transmitted with the position data. You do not have to set 'Brand' to Motorola, but you can if you want to.

I'm not sure about the APRS interval. Changing the setting in Brandmeister seems to have no effect. It does not matter if I have beacons turned on or I am just using the PTT method, only the radio timing settings seem to matter.

APRS options

1. GPS must be turned on and be locked (red splodge) on the radio display.

2. You can configure APRS to send location data on one of the APRS talk groups when you release the PTT. Set **Manual TX Interval** to at least 60 seconds to avoid sending APRS data every time you press the PTT button.

3. Or you can configure the radio so that it sends an APRS beacon every few minutes. The delay is adjustable from a beacon every 30 seconds, to one every 7650 seconds (2.125 hours).

4. If you leave the Report Channel set to Current Channel and add the 234999 or 310999 talk group You can make the radio, send APRS data on every channel, provided you also set the APRS settings on each channel. I strongly discourage this. It is just not necessary. Also, a caller may give up waiting while your APRS packet is being sent after you stop talking.

5. You can choose a specific channel, or channels, for sending APRS as well as speech. I think this is OK for sending the occasional APRS packet while hiking. But it does trigger the repeater every time a beacon is sent, so it could annoy other users if you set the beacon period too short, or if you decide to transmit beacons for an extended period. Especially if you are not travelling anywhere.

6. You can set up a channel especially for APRS with the talk group set to 234999 or 310999. You probably can't use it to talk to anyone. But it works very well when you are using the 'release the PTT method. Just 'kerchunk' the PTT and away goes the APRS beacon.

Summary of the process

1. Create a talk group for APRS on TG 234999 if you live in the UK, or TG 310999 if you live in the USA, or TG 5057 if you are on the UK Phoenix DMR Network or TG 9057 for DV Scotland. If you live anywhere else and you have a Brandmeister repeater or hotspot you can use either TG 234999 or TG 310999, they both work fine.

2. Set up the APRS tab under **Common Setting > APRS**. If the APRS page is not there it means you have not turned on the option **Tool > Options >APRS**.

3. Optionally set up a channel for APRS or change the APRS settings in the nominated channel.

4. Trigger the radio PTT or wait for the beacon timer. A message on the radio tells you that APRS data is being sent.

5. Go to APRS.fi and see your location plotted on the map.

APRS ON TYT RADIOS

You can set up Brandmeister to poll your TYT radio's position regularly and forward the location as an APRS beacon to APRS.fi, for plotting on the map. You will need to set up a channel to turn on the process and another to turn it off. Or you can turn it off by setting APRS Interval to Off on the Brandmeister Self help page. This is the method described by K2GOG http://omarcclub.org/forum/discussion/243/using-gps-on-tyt-md-380.

1. Turn on the GPS receiver in the CPS.

2. Make a new **private call** talk group in the Digital Contacts list for **TG 310999,** or **TG 234999** if you live in Europe. Call it **APRS 310999** or **APRS 234999.**

3. Create a new channel on your repeater or hotspot. If it is a repeater channel, make sure that the repeater owner allows APRS. A lot do not. Call the new channel **APRS USA On** or **APRS UK On** and set the **contact name** to the talk group you created in the last step. Use the hotspot's **colour code** and **TS2.** Check the **Send GPS Info** and **Receive GPS Info** boxes. Set GPS system to **1.** (You can use TS1 on a duplex hotspot).

4. Create a second channel the same as the one above, except call it **APRS USA Off** or **APRS UK Off.** Set GPS system to **None.**

5. Add both channels to an existing or new **Zone.**

6. Go to the **GPS Settings > GPS1** in the CPS and set **GPS Revert Channel** to the channel you created for **APRS USA On** or **APRS UK On.**

7. Set your default **GPS report interval** to a value of **180.** If you select a smaller value, the GPS constantly transmitting will affect your battery life.

8. Set the **Destination ID** to the Digital Contact you created **APRS 310999** or **APRS 234999.**

9. Save the code plug and download it to the radio

Confirm that the radio is seeing the GPS satellites

You may have to wait 10 minutes for the radio to achieve a GPS lock. You may have to take the radio outside. On the radio select **Utilities > GPS BeiDou Info.** If the radio is receiving signals from the GPS satellites, you should see your latitude, longitude, altitude and the number of GPS satellites your radio can "see."

Brandmeister setup

Login to Brandmeister and open the **Self Care** page (click your callsign at the top of the page). Select **Chinese Radio,** enable a **60-second APRS interval** and select the appropriate SSID for your call sign and icon. It can be just your callsign or callsign and extension. Brandmeister will poll your radio through your hotspot (or repeater) and read the GPS data. It will format it for APRS and send it to APRS.fi as a 'beacon.'

To start APRS uploads, Key Up your **APRS USA On** or **APRS UK On** channel and to stop it, Key Up your **APRS USA Off** or **APRS UK Off** channel. Or you can set the **APRS interval** to **Off** in Brandmeister.

Go to APRS.fi and enter your callsign into the search box. If the APRS data has been received, the map should re-centre on your location and display the APRS icon. Click on that to check the time that the beacon was received.

Finding repeater information

There are several ways to find out what static talk groups are available on your local repeater. You also need to know the repeater frequencies or transmit frequency and offset, and its colour code, for setting up your channels. I believe that the easiest way is to look at the IPSC2 or Brandmeister dashboard, depending on what DMR network is being used. You could also consult the website for the repeater (if there is one), or the club that owns the repeater, or a relevant national or regional DMR group.

If the repeater is on the Brandmeister network, you can look it up on https://brandmeister.network. Click 'Repeaters' then enter the repeater callsign or partial callsign (e.g. KH6) into the Search box. If the repeater is listed, it is connected to the Brandmeister network. Click the blue repeater callsign to display the information page. Frequency details and static time slots are listed on the left side. Take a careful note of the time slots for the static time slots. You will need that information when you set up channels for accessing the repeater.

If the repeater is not on the Brandmeister network but the network has a public IPSC2 dashboard, you can use that to find out about the static talk groups on the connected repeaters. The Matrix tab on the IPSC2 dashboard shows the static talk groups for TS1 and TS2. The first table is for TS1, and the second table is for TS2.

RPTR bMaster+	ID	INFO connected	8/0	153 **
ZL2R	530009	Wellington	X	
ZL3DVR	530305	MultiMode Repeater	X	X
ZL4GQ	530015	Invercargill	X	X
ZL4IND	530306	Queenstown		X

Figure 39: IPSC2 Matrix tab TS1 static talk groups

The IPSC2 'Status' page also lists the static talk groups in the TS1 and TS2 columns. The 'Service' page shows repeater frequencies, callsign, static talk groups, reflectors, and time out timer.

If the Pi-Star dashboard is public, a Google search of the repeater callsign should find it. The Pi-Star dashboard shows the repeater status and its frequencies, but not the static talk groups. The information is there, but not publicly accessible.

Another method is to look the repeater up on repeaterbook.com. Click DMR Tools then select Worldwide or North American repeaters on the green bar, then your country or state. In the 'Feature' section click DMR and you will get a list of DMR repeaters in your country. Clicking the blue frequency hyperlink brings up information about the repeater including the frequencies and sometimes the static talk groups. The information that is provided varies. Many of the listings include a link to a website specifically for the repeater or network.

IPSC2 dashboards (DMR+)

The IPSC2 dashboards show all the connected repeaters and hotspots for DMR+ and connected networks. Hotspots may have a separate webpage. You can see who is transmitting and what talk group and time slot is in use. There will be an IPSC2 dashboard for your regional master server. The Brandmeister network uses different linking protocols. See the next chapter.

Each repeater's static talk groups are listed on the IPSC2 dashboard 'Status' page in the 5th column for TS1 and the 8th column for TS2. The top section shows any active transmissions, the bottom section shows connected repeaters or hotspots. When a station transmits, all of the repeater lines with that talk group change colour to indicate that they are transmitting the data.

IPSC2 stands for Internet Protocol Site Connect. It is the networking protocol that is used to link regional DMR systems and DMR+ repeaters and hotspots to the Master Servers and from there to the world. It also interfaces with c-Bridges, OpenBridge for connecting to the Brandmeister network, and Interlink, for connecting to other IPSC2 servers. In short, IPSC2 joins the DMR+ network together.

IPSC2 DASHBOARD - TABS

Status – displays all of the connected repeaters, hotspots, Interlinks, and OpenBridge connections. This is the most useful page.

Monitor – displays roaming data and the 'last heard' list. Current calls are shown in the top section. You can see who has been using the DMR+ network.

Matrix – displays the static talk groups for each repeater or hotspot. The first table is for TS1, and the second table is for TS2.

Remap – shows Reflector to Talk Group connections, (TG, TS, Reflector).

Bridge – shows talk groups and time slots that have at least one static link (green), or dynamic link (orange). A grey link means that the talk group is available but not in use on the IPSC2 network, so there is no global routing for it. It can indicate a regional talk group. 'Not Booked' relates to static talk groups that are not currently in use. 'Stand By' is for dynamic talk groups that are available but not in use.

Dongle – shows linked DMR dongles. I have never seen anything on this page.

Service – shows the hardware, repeater ID, callsign, frequency, repeater offset (if duplex), static talk groups for TS1 and TS2, home reflector, time out in minutes, firmware version, and a serial code, for each repeater. Useful for finding out the frequency and static talk groups on your local repeaters.

Site Info – shows an information page and the contact details for the site administrator.

IPSC2 DASHBOARD – STATUS PAGE

The Status tab displays all connected repeaters, hotspots, Interlinks and OpenBridge connections. The top section shows active links to reflectors. The bottom section shows time slots being used on each of the connected repeaters.

NR	REPEATER	INFO	ID	TS1	CQ	LINK-STATUS TS1-INFO	TS2
1	GB3GS	Downham Market (28)	234107	235	CQ		840
2	GB3OM	Omagh (29)	235501	1 2 13 235	CQ		8/1 880
3	GB7AB	Aberdeen (30)	235488	2351	CQ		
4	GB7AL	Ipswich (27)	235153	1 2 13 235	CQ		840
5	GB7AS	AshfordKent (29)	235148	1 2 13 235	CQ		801
6	GB7AV	Aylesbury (18)	235116	1 2 13 235	CQ		810 842
7	GB7BA	Bradford (29)	235214	1 2 13 235	CQ		

Figure 40: IPSC2 status - lower section

The 'repeater' column shows the repeater callsign and the 'Info' column shows the repeater's identity. The number in brackets that keep counting down from 30 to about 20 is the 'watchdog timer.' The repeater must respond to the server within 30 seconds, or it will be removed from the status page. You will see a similar timer when a station finishes with a time slot. That counter shows the time before the linking stops. The ID column shows the DMR ID of the repeater or hotspot.

The numbers in the TS columns show the static talk groups assigned to the repeater. For example, line two on the image above shows the Omagh GB3OM repeater. TS1 has static connections for TG 1, 2, 13, and 235. TS2 has static connections for TG 8 or 1, and 880. A number in brackets in the TS column indicates a dynamic link to a non-static talk group has been established by a user pressing the PTT on their radio. The number beside it will count down from 15 minutes. If there is no further RF activity on the repeater to reset the counter, the repeater will return to the static talk group(s).

TS1 and TS2 operate independently on each repeater. A coloured bar in the TS1-INFO or TS2-INFO column means that the timeslot is active on that time slot.

A **green bar** indicates that the repeater is carrying the station that initiated the call by pressing the PTT on their radio. If the call came in from another server, the green bar will be on CC link, YCS link, Interlink etc. at the bottom of the page. The text will show the TG and possibly the TG name, plus the DMR ID of the originator and their callsign.

96	CAN-LINK	InterLink (26)	183271	3 3172 8801 302 320...	320/USA (3133987/WX4QZ)

Any repeater that has a static, or an active dynamic, talk group that matches the talk group selected by the instigator of the call, will show up as an **orange bar.** The repeater will transmit the talk group data.

The orange box also indicates the TG, TG name, the DMR ID of the originator, and their callsign.

If a popular talk group like 3125 is selected, hundreds of repeaters on the IPSC2 server and possibly thousands of repeaters worldwide will repeat the call. Every repeater that has 3125 as a static talk group will 'go orange.'

You sometimes see a **red bar** on a repeater time slot. This happens on non-duplex hotspots when it receives data for one time slot when it is already transmitting on the other time slot.

IPSC2 DASHBOARD – PAGE FOOTER

Interlink:2	cBridge:1	Motorola:0	Hytera:0	MMDVM:9	HOTSPOT:32	Dongle:0

Max User DB:142708	bMaster+:89.185.97.34	Starttime: 2011-11-18 22:41:59	ID 397624

Figure 41: IPSC2 webpage footer

The footer at the bottom of the IPSC2 website indicates,

- The number of Interlink connections between the displayed IPSC2 server and other IPSC2 servers.

- The number of c-Bridge CC-CC (control centre - control centre) connections

- The number of Motorola repeaters connected (different data protocol)

- The number of Hytera repeaters connected (different data protocol)

- The number of MMDVM repeaters connected

- The number of MMDVM hotspots connected

- The number of DMR dongles connected

- Max User database

- The URL of the master server

- The 'up-time' of the IPSC2 server, and

- Its DMR ID.

Brandmeister dashboard

The Brandmeister network uses 47 master servers. It does not use IPSC2 dashboards, but you can see the server connections, hotspots, and repeaters at https://brandmeister.network/. Click on repeaters or hotspots and enter the repeater callsign (or partial callsign) into the search box.

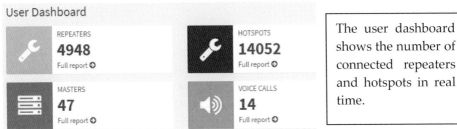

The user dashboard shows the number of connected repeaters and hotspots in real time.

Figure 42: Brandmeister dashboard

The map can be zoomed. It shows active talk groups in real time.

Figure 43: Activity map

Number		Name		Hardware		Firmware
223352		IW3GPO2		MMDVM (Repeater)		20210617_PS4
223360		IW3IHV-1-VERONA		Motorola SLR5500		R02.09.00.13
223371		IW3GXX-1		MMDVM_HS_Dual_Hat (MMDVM)		20181107_Pi-Star

Search the repeater display for a name, callsign, frequency, master server, hardware type, firmware revision, or DMR ID. You can do a partial search like 'ZL3' or 'Pi-Star.' Most entries include a national flag. Clicking a callsign opens an information screen relating to the station.

The Brandmeister dashboard services all Brandmeister talk groups, repeaters and hotspots worldwide. It has a lot of features, and I don't have space to mention them all. The following section is an overview of what is available. As well as the Dashboard there is a Brandmeister Wikipedia page, but it is not a complete guide. There are some hidden pages on the website such as the one showing Brandmeister talk groups. I found that by using Google.

USER DASHBOARD

Network status

The top part of the user dashboard has three sections. At the top left, there is a real-time display of the network status. (See the image above). It shows connected repeaters, master servers, hotspots, and the current number of voice calls on the system. It filters out 'kerchunks.' I was surprised how active voice calls few there are. You can expand all of the sections for more information. You get diverted to the same page as you get by selecting the same item on the black menu bar. Clicking **Voice** takes you to the 'Last Heard' page.

Repeater status

Below the network status icons, there are three 'wheels' which show the number of repeaters currently receiving a signal, the number that are transmitting, and external links.

Map

The map shows active talk groups. If a USA wide talk group activates, the entire country is highlighted. Including Hawaii and Alaska. The same with Russia, China, Europe, and Australia.

Charts

If you scroll right down the page, you will see some charts showing connected repeaters and hotspots, repeater models (dominated by MMDVM hotspots) and firmware versions. Click on the pie chart to reveal even more information.

REPEATERS

If you have a hotspot or a repeater connected to the Brandmeister network you can log on to the site and edit details, especially the static talk groups. If you have a duplex hotspot the site treats it like a repeater.

Clicking on the **Repeater** icon or menu option opens the 'Repeaters' page. It will be blank for several seconds as the list currently contains 5002 entries. The page is best used with the search function.

You can enter a full repeater callsign or a partial such as KH6 or ZL3. You can enter Motorola or Hytera or MMDVM, or a frequency, or a DMR ID (including partials). The search engine does not support wildcards, just partial entries. Entering 530 brings up all DMR IDS that contain 530, not just ZLs. ZL brings up all callsigns that contain ZL, so it is better to enter ZL3 or ZL2 etc.

Clicking on a header changes the way the data is indexed.

Status: 'Slot 1 & 2 linked' does not mean that the time slots are linked together. It means that both time slots are linked to the Brandmeister network. Some repeater owners only link one time slot to Brandmeister.

Repeater information file

Clicking on the blue repeater name takes you to the information file for that repeater. This shows a heap of information. A map showing the repeater location, frequencies, colour code, static talk groups hardware and firmware, linked time slots (some repeaters only link one time slot to Brandmeister and use the other for a local DMR network. There is also a 'last heard' list showing activity and signal quality. There are no secrets on the Brandmeister network!

The time slot details show static talk groups, dynamic talk groups, cluster connections, and static talk groups that activate on a timer.

HOTSPOTS

Clicking on the Hotspot icon or menu option opens the 'Hotspots' page. It will be blank for several seconds as the list currently contains 13,882 entries. The page is best used with the search function.

'Hotspots' lists simplex hotspots only. Duplex hotspots are listed as 'Repeaters.'

The search functions are the same as the repeater page. You can enter a full callsign or a partial such as KH6 or ZL3. You can enter a callsign, frequency, or DMR ID (including partials). The search engine does not support wildcards, just partial entries. Entering 530 brings up all DMR IDs that contain 530 and frequencies that include 530, not just ZLs. Try entering 5300 instead. KH brings up all callsigns that contain KH, so it is better to enter KH6 or KH7 etc. A four-digit number could be a Brandmeister master server number or part of a DMR ID.

Clicking on a header changes the way the data is indexed.

Status: you will see that all entries are shown as 'Linked in DMO mode.' Simplex hotspots are linked in DMO (Direct Mode Operation) mode. It means the mode which two DMR radios use to talk directly radio to radio. In other words, it is a fancy way of saying "Simplex operation with a network connection."

Hotspot information file

Clicking on the blue hotspot name takes you to the information file for that hotspot. The page is identical to the repeater information page, but there may be less information provided than a typical repeater record. A map showing the hotspot location, simplex frequency, colour code, static talk groups, hardware and firmware. There is also a 'last heard' list showing activity and signal quality. There are no secrets on the Brandmeister network!

The time slot details show static talk groups, dynamic talk groups, cluster connections, and static talk groups that activate on a timer.

MASTERS

The 'masters' list shows the status of the 47 master servers, including the country that they are located in. The MS number conforms to the MCC code plus a 1 or a 2. For example, Australia is 5051.

The information on this screen is for the DMR network operators and 'real' repeater owners and is of little interest to anyone else.

Status provides up to date linking status and is of no interest to most folks.

List shows connected repeaters and hotspots, including the callsigns and DMR ID. If the status is reported as 'Linked in DMO mode' the connection is to a simplex hotspot. If it says, 'Slot 1 & 2 linked' the connection is to a repeater or a duplex hotspot.

Monitor shows more repeater information including telemetry.

System shows complicated IP networking information

ALERTS

The Alerts page is for reporting repeater environmental alarms such as VSWR or Low Battery. It is for 'real' repeater owners and is of little interest to anyone else. I guess you could check this page if your local repeater is off the air, and you want to know why.

DATA VISUALISATION

Talk groups

Use the search box to list the talk group(s) you are interested in.

You can search for talk activity on any Brandmeister talk group. Enter the talk group number or a partial match. For example, entering 530 brings up 23530 as well as the New Zealand talk groups. Entering NZ brings up talk groups that happen to have NZ in their name, and ZL only brings up talk groups with ZL in the name. As usual, you can index the results by clicking the column headers.

- **LH** opens the Last Heard screen for the selected talk group and that is searchable as well.

- The **Hoseline** feature lets you hear the activity on a talk group. Click Player (top right) to add more talk groups and see the source DMR ID, Destination talk group, and Talker Alias (name and callsign) if it is turned on.

- **Wiki** references the Brandmeister Wikipedia page to the selected talk group.

Clusters

The 'Clusters' page shows active clusters, with the usual search parameters based on the name or country. You can click the 'Show members' button to display repeater information.

Network map

The network map takes a while to populate. It shows the number of Brandmeister repeaters worldwide. You can zoom into the individual repeater locations. Clicking a repeater icon opens a 'Maps' information box, usually containing the repeater's callsign, frequencies, colour code and a link to 'last heard.'

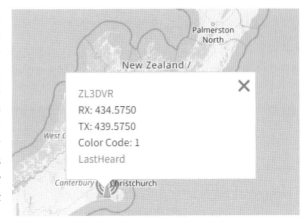

Call statistics

The call statistics page shows you a plot of QSOs per hour and 'Time in calls' per hour. You can click and drag to highlight any time or date range. Enter a talk group using the talk group dropdown box, to view the statistics for a particular talk group. You can turn the talk group display on and off with the switch beside the dropdown box.

Contacts export

This is an interesting page. You can use it to get a .csv contacts file, relating to the users of a talk group or several talk groups. Enter the talk groups and press the **Run** button. There seems to be about seventeen months of records. Back to June 2020 at the moment. Click **LastTX** to sort the list in date and time order.

Click **TX Count** (twice) to see who uses the channel the most. And you can search for a callsign, a name, or a DMR ID.

INFORMATION

Takes you to the BM Wiki pages.

LAST HEARD

The Last Heard menu option takes you to the Last heard web page, which displays the most recent 30 calls in real-time. The pause symbol above the small search box freezes the display. You can perform the usual search operations using the search box.

TIP: If you want to search for a talk group include the brackets or at least the opening bracket. It stops you from getting talk groups and DMR IDs that contain the same number. For example, searching on 91 will bring up 5300391. Searching on (91) will only bring up talk group 91. Searching on (310 will bring up all talk groups that start with 310 and all DMR IDs that start with 310.

If you are really keen you can click the + symbol at the top right of the screen and enter your more complex search parameters. Then click **Search**. Here are some examples.

Add Rule > AND > Source ID > Contains 2349 will only show callsigns with a DMR ID that includes 2349.

Add Rule > AND > Destination ID > Equal 530 will only show calls to talk group ZL National 530.

Add Rule > AND > My Call > Contains ZL3DW show calls made by ZL3DW.

The headers along the top can be used to index the list.

Time is the age of the record

Link name is the type of equipment, MMDVM for a hotspot, Motorola or Hytera for repeaters, YSF for a link to a System Fusion network, C-Bridge, Open Bridge etc.

My Call is the originating callsign [name] and (DMR ID)

Talker Alias – name and callsign if sent by the radio, or the hotspot

Source is the repeater callsign. If the originator is using a hotspot, it will be the same callsign and DMR ID as 'My Call.'

Destination is the talk group that the originator is using.

Options usually indicates DMR but can show a YSF or D-Star user accessing the DMR network through a linked talk group.

RSSI is the signal strength into the repeater or hotspot if reported. S9+40dB indicates that you are transmitting close to your hotspot. An indication less than S9 usually suggests a call into a repeater.

dBm This is the received signal strength in dBm instead of S points.

Duration is the call duration in seconds

Loss rate is the loss rate reported by the repeater as a percentage and as a ratio

SERVICES

The Services option provides two useful choices. Hoseline and SelfCare.

Hoseline

Hoseline enables you to hear conversations on the active talk groups. You can select more than one. The page opens with a display of all the current voice calls. Talk groups with an orange border are carrying live speech. Those without a border were carrying speech recently. The counter counts up while a QSO is happening and down to zero after it stops. Click on the orange symbol in the top right of each active box (orange border) to hear the traffic.

Player

In the top right of the window, there is a 'Player' icon that opens the audio monitor. There are stop and play icons at the top of the popup box, followed by any selected talk groups. You can use the dropdown list to select a talk group or click on the orange symbol in the top right of any active talk group box (orange border) to hear the traffic. You can monitor several talk groups at the same time. The one you hear is "first in first served."

'Src' is the source DMR ID and 'Dest' is the TG number. TA is the Talker Alias, which will display a callsign and name if TA is active on the radio or the hotspot.

Player signal strength meter

The white line below the talk groups is a signal meter. Hover the mouse to see the signal level and the peak level in dBm. This function is very handy. If you monitor TG 98 in Hoseline Player and make a 'test call' on that talk group, you can monitor your transmission. Set the radio's microphone level for a signal that just peaks to green on the signal strength meter. You will see the strength that other stations are using. Your signal should never reach the red indicator. It will become distorted at that point.

Self-Care

Self-Care requires you to log in to the site. It is where you set your details, password, APRS text, APRS beacon etc.

Once you have logged in, your callsign should appear at the top of the web page. Click that to edit your profile, upload a photo, change your profile settings, or return to the Self-Care page.

MY HOTSPOTS / MY REPEATERS

Login required. Clicking the My Hotspots or My Repeaters menu option opens a dropdown list showing your hotspot(s) or repeater(s). Clicking that opens the setup page for the selected hotspot (or repeater).

*TIP: Click the (**View**) link next to the 'Settings of Callsign' at the top of the web page if you want to view the page that people will see if they click your callsign in the Brandmeister repeater and hotspot lists.*

General Settings

The **Priority Message** and **Description** boxes have something to do with APRS. Entering NOAPRS in the priority message box appears to block APRS from Brandmeister but does not inhibit APRS beacons sent from your radio.

The **Website** is usually your QRZ.com listing, where people can look up information about you and your station. You can't change it here. It is set in the Configuration of your Pi-Star dashboard.

Location (City) is for your city or location plus your Maidenhead grid. 'Christchurch, RE66hm.' The page also displays your Latitude, Longitude and Height above ground level (AGL). You can't change these items here. They are set in the Configuration of your Pi-Star dashboard.

You can edit the **Power ERP** level using the up-down arrows on the text box. But the minimum is one watt which is much higher than most hotspots, so this is really for repeaters. ERP is effective radiated power. It takes into account the loss of the feeder cable and the gain of the antenna.

Sysops

The Sysops (system operator) area is usually unpopulated. You always have full Sysops control. This area is for you to add a second person. You would probably only do that if the repeater or hotspot was owned by a radio club or other group.

Actions

This area manages the connection between Brandmeister and your repeater or hotspot.

- **Get IP address**, reports the IP address of your hotspot or repeater. This is not the IP address you use inside your LAN (for Pi-Star) etc. It is the IP address that Brandmeister uses to access hotspot or repeater. It is also the IP address that people would use to access your Pi-Star hotspot over the Internet if you elected to give it public access.

- **Drop call on slot 1**, drops the talk group being sent to your repeater or hotspot on TS1. The reason for this is that a talk group might be held up by a faulty radio setup, or an idiot on the channel. If the time slot is busy you (and other repeater users) are unable to use that time slot on a different talk group.

- **Drop call on slot 2**, as above. It drops the talk group being sent to your repeater or hotspot on TS2.

- **Drop dynamic groups on slot 1** and **drop dynamic groups on slot 2**, removes the dynamic talk groups and drops any traffic being sent to the hotspot or repeater. This is done for the same reason as dropping a call on a time slot.

- **Reset repeater connection**. Drops then re-establishes the network connection to your hotspot or repeater.

Static talk groups

This is where you set your static talk groups. With a duplex hotspot or a repeater, you can allocate talk groups on both time slots. With a simplex hotspot, you can only allocate talk groups on time slot two.

Enter a talk group number in the left box and click the arrow → to put it into the list. You will be rewarded with **Success**. Some talk groups cannot be set as static. Notably the US TAC channels. The left-facing arrow ← moves a talk group out of the list.

TIP: If you are just starting out, try setting TG 91 to static. It is very busy, and you should see activity within five minutes at any time of the day or night.

TIP: I only set static talk groups on TS1 because I use TS2 for the DMR+ network.

Clusters You can add a cluster as static.

Scheduled talk groups

You can schedule talk groups so that they become static at certain times of the day, or on certain days. This might be useful if there is some sort of short term event on a talk group such as JOTA. I would probably just make it static and remove it after the event, but you might find the option useful. For example, you might set a talk group to static for a Net that happens at a particular time and day of the week. The rest of the time the channel is very busy and you don't want the talk group to be static.

TGIF dashboard

The TGIF dashboard at https://tgif.network/index.php shows the status of the network and offers some other interesting features. It operates more like the Brandmeister dashboard than an IPSC2 server dashboard, being a full website. TGIF operates through a single server, so all the network information is in one place. The website states that there are currently 5804 registered users. There is a link to 'Discord' a third party "chat page" which supports voice calls. I can't see the point of using an Internet app for voice calls between TGIF subscribers when you can use the DMR network.

MONITORING

The **Last Heard** page shows the calls on the network in real-time. It can be filtered by entering information into the Search box. Enter a partial or full callsign, a talk group number, or a first name. Even a time can be entered. You can also sort the list on any column. Selecting 'Timestamp' descending puts the most recent calls at the top of the list. TGIF does not provide a Hoseline Player so you cannot hear the QSOs.

TIP: I find the Last Heard page useful to see what talk groups are being used. I can then Key Up or select a busy talk group in the Self Help area so that I can hear the traffic on my radio. If you don't have the talk group loaded, you can use the digital monitor feature on the radio.

The **Active Talk Groups** screen shows recent activity on talk groups that are currently linked to hotspots. If you recently keyed up a channel, you should be on the list. A blue icon with a white callsign and first name indicates that the station is linked to the talk group. Click on the link to show the station details. A grey icon with no name indicates that the station is linked but there is no station information. A green icon indicates that there is voice traffic, or the station is 'keyed up' on the talk group.

TIP: A good way to see if you are reaching the network is to key up on a TGIF channel and watch it come up on the Active Talk Groups display.

The **Talk Group List** provides a way to download the TGIF talk groups as a .csv or .json file. You can search the file and sort it on any column. Clicking a talk group name takes you to an information page explaining what the talk group is used for and contact information for the talk group trustee. Try TG 119, TG 404, or TG 31690.

The **Server Status** page has information about the TGIF network. It is of little interest unless the network is down for some reason.

HELP

The help page provides setup instructions, lodging a trouble ticket, a user forum, FAQ, and links to some good videos.

CALLSIGN LINK

Assuming that you have registered and are logged on, the link (top right) with your callsign on it has settings relating specifically to you.

Change Theme changes the way the website looks.

Profile settings lets you change the information you submitted when you registered. And lets you change your callsign.

The five icons at the top left of the page are quick links for; edit profile settings, change **email address**, change website access password, self-care page, and user security. I think that this is the only way to change your email address.

User Security is an important page. It provides a security key that you should use in your Pi-Star or another hotspot setup. If for some reason the supplied randomly generated code doesn't, please you, click the green key for a different one. *Don't click this option if you have already used the security key in your hotspot setup.*

Click the blue key to copy the number to your PC clipboard (or use CTRL+C), then paste it into the password line on the hotspot setup page.

TIP: Some TGIF talk groups will work with the default 'passw0rd' but increasingly, access to the busier talk groups requires this 'registered' security key. Paste it in now and you don't have to worry.

The **Selfcare** page is used to set auto-static talk groups. It displays information about your IP address and port, hotspot, and frequencies. Clicking the blue button with your DMR ID and callsign on it pops up an information window.

The blue lozenge(s) show the auto-static TS1 and TS2 talk groups, (a simplex hotspot has TS2 only). You can change them (duplex) or it (simplex) here or by keying up a different talk group on your radio. The choices seem to stick until you change them.

Click on the blue lozenge. Enter a talk group **name** or **number** in the 'ID or Name' box or make a selection from the dropdown list and click **Update**.

TIP: The TGIF network does not use TG 4000 disconnect. Just key up a different talk group.

It is much easier to establish a talk group for your special interest, club, or activity on the TGIF network than on the other DMR networks. Any registered TGIF user can **Request** a **New Talk Group**. I don't recommend doing it unless you have a genuine reason, such as a group that wants a place to meet. There is no point sitting in a room on your own!

Finally, there is the **Logout** option. You don't have to log out. But you can.

Pi-Star Dashboard

This chapter covers the features of the Pi-Star dashboard after you have configured it for the hotspot and DMR network(s) you are using. There is a lot of information presented on the main dashboard page. Click **Dashboard**.

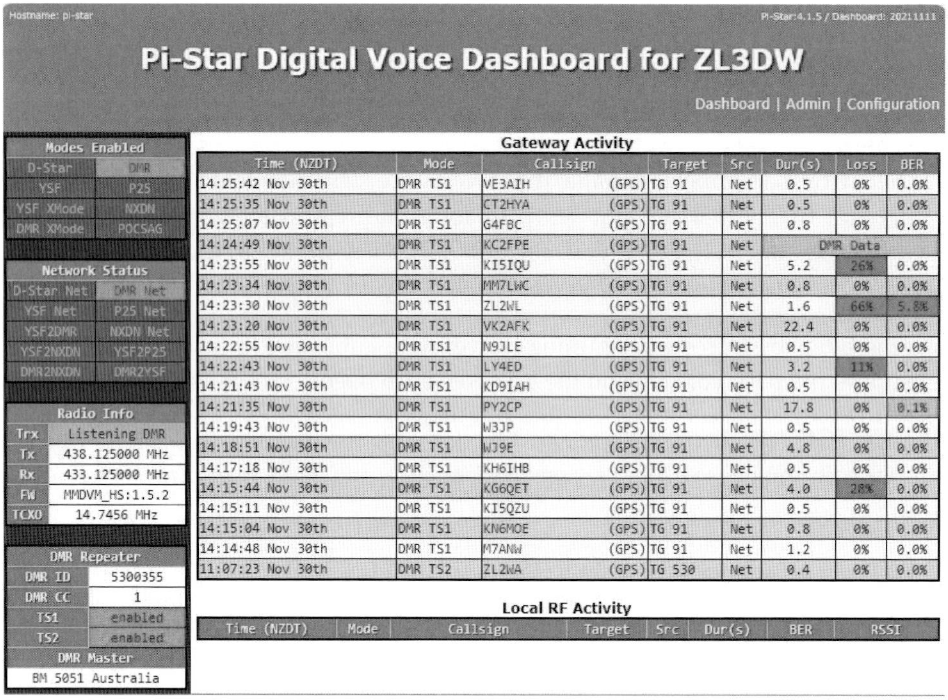

Time (NZDT)	Mode	Callsign	Target	Src	Dur(s)	Loss	BER
14:25:42 Nov 30th	DMR TS1	VE3AIH (GPS)	TG 91	Net	0.5	0%	0.0%
14:25:35 Nov 30th	DMR TS1	CT2HYA (GPS)	TG 91	Net	0.5	0%	0.0%
14:25:07 Nov 30th	DMR TS1	G4FBC (GPS)	TG 91	Net	0.8	0%	0.0%
14:24:49 Nov 30th	DMR TS1	KC2FPE (GPS)	TG 91	Net	DMR Data		
14:23:55 Nov 30th	DMR TS1	KI5IQU (GPS)	TG 91	Net	5.2	26%	0.0%
14:23:34 Nov 30th	DMR TS1	MM7LWC (GPS)	TG 91	Net	0.8	0%	0.0%
14:23:30 Nov 30th	DMR TS1	ZL2WL (GPS)	TG 91	Net	1.6	66%	5.8%
14:23:20 Nov 30th	DMR TS1	VK2AFK (GPS)	TG 91	Net	22.4	0%	0.0%
14:22:55 Nov 30th	DMR TS1	N9JLE (GPS)	TG 91	Net	0.5	0%	0.0%
14:22:43 Nov 30th	DMR TS1	LY4ED (GPS)	TG 91	Net	3.2	11%	0.0%
14:21:43 Nov 30th	DMR TS1	KD9IAH (GPS)	TG 91	Net	0.5	0%	0.0%
14:21:35 Nov 30th	DMR TS1	PY2CP (GPS)	TG 91	Net	17.8	0%	0.1%
14:19:43 Nov 30th	DMR TS1	W3JP (GPS)	TG 91	Net	0.5	0%	0.0%
14:18:51 Nov 30th	DMR TS1	WJ9E (GPS)	TG 91	Net	4.8	0%	0.0%
14:17:18 Nov 30th	DMR TS1	KH6IHB (GPS)	TG 91	Net	0.5	0%	0.0%
14:15:44 Nov 30th	DMR TS1	KG6QET (GPS)	TG 91	Net	4.0	28%	0.0%
14:15:11 Nov 30th	DMR TS1	KI5QZU (GPS)	TG 91	Net	0.5	0%	0.0%
14:15:04 Nov 30th	DMR TS1	KN6MOE (GPS)	TG 91	Net	0.8	0%	0.0%
14:14:48 Nov 30th	DMR TS1	M7ANW (GPS)	TG 91	Net	1.2	0%	0.0%
11:07:23 Nov 30th	DMR TS2	ZL2WA (GPS)	TG 530	Net	0.4	0%	0.0%

Time (NZDT)	Mode	Callsign	Target	Src	Dur(s)	BER	RSSI

Figure 44: Pi-Star Dashboard page

Modes Enabled shows the modes that you selected in the Configuration screen. This book only covers the DMR selection. If DMR is red, it means that it has become unselected.

Network Status. DMR Net (green) indicates that the Internet connection is up, and the system is communicating with at least one DMR network. There is no indication if your second or third network disappears.

Radio Info shows the status of the hotspot.

Trx

- Listening DMR (orange) means that the user has stopped transmitting and the hotspot is waiting for the network hang time to expire. If no signal is heard during the hang time the hotspot will go to its idle condition allowing other talk groups to access the time slot.

- TX DMR Slot 1 and/or Slot 2 (red), means that the hotspot is transmitting on the time slot indicated.

- Listening (green) the hotspot is in its idle condition waiting for a call to arrive from the DMR network or the RF path.

TX and **RX** show the hotspot's transmit and receive frequencies, set in the configuration.

FW is the firmware revision of the hotspot.

TCXO is the nominal oscillator frequency. This is what you adjust if you apply an offset. They usually run at either 12.288 MHz or 14.7456 MHz.

DMR Repeater has information about the hotspot.

- DMR ID is your DMR ID

- DMR CC is the hotspot colour code (for all channels)

- TS1 enabled means that TS1 is turned on (duplex only)

- TS2 enabled means that TS2 is turned on

DMR Master shows your network connection(s). The image above shows a connection to Brandmeister server 5051 in Australia. I have now added DMR+ IPSC2 NZ Hotspot and the TGIF network.

GATEWAY ACTIVITY

The gateway activity area shows calls broadcast by the hotspot that originated on the Internet (network) side. These are indicated with 'Net' in the Source column. Calls with 'RF' in the Source column originated as an RF signal received by the hotspot. Usually, that will only be your handheld or mobile radio(s). If the Pi-Star dashboard is for a repeater, there could be many RF calls.

Other information includes the date and time of the call, the mode and time slot, the callsign, the talk group that was called, the source, the duration of the call, the amount of data loss (after forward error correction) and the BER (bit error rate) quality of the signal. Clicking on the callsign takes you to the user's nominated website. Usually but not always their QRZ.com listing. Note that (GPS) does not mean that the radio is equipped with GPS. It is a link to the APRS.fi website with the callsign loaded as the search quantity. A 'DMR data' (green bar) indicates a data call, usually APRS data.

LOCAL RF ACTIVITY

The local RF activity area provides pretty much the same information for calls that originated on the RF side of the hotspot. Plus, the received signal strength. For a hotspot on your desk, it is usually massive, S9 +46dB (-47 dBm). It is more useful for measuring the signal from stations accessing a DMR repeater.

THE ADMIN PAGE

The 'Admin' page adds **Active Brandmeister Connections** if you are connected to the Brandmeister network, listing your static and any dynamic talk groups. It also shows the Brandmeister server (twice) and the BM repeater ID including any ESSID extension. *TIP: The access username is pi-star, the password is raspberry.*

At the top of the page, there is **Gateway Hardware Information** about the Raspberry Pi. It shows the Host Name (which we decided not to change), The Pi Kernel release, The Pi model, its CPU loading per core, and the CPU temperature. If it gets into 'the red' you need to add a heatsink or fan to improve the Raspberry Pi CPU cooling. Mine runs into orange at 50.5 degrees Celsius, but it seems to be stable, and I have a heatsink on it.

I don't know about the **Service Status** section. I guess green is good.

Even though it is the "Admin" page, you can't change any settings. But it does add three more menu items. Live Logs (no idea), Power – to shut down or reboot the Pi, Update - only click this if you want to do an update. You won't get asked again!

THE CONFIGURATION PAGE

We will get into the configuration settings later. The configuration page does add some more menu items. Expert – for experts like us, Power - shut down or reboot the Pi, Update - only click this if you want to do an update. You won't get asked again! Backup/Restore, and Factory Reset. Don't click that or you will erase all your configuration settings.

Backup/Restore – you should make a backup as soon as your configuration is stable and working. Backups are downloaded to your standard Windows download directory as a Zip file. You can leave them there or move them to your CPS directory.

The zip file contains twenty configuration files. Editing settings in the 'Expert' area changes the relevant config file. Most are text files that can be viewed or edited (if you are especially brave) in Notepad or Wordpad. I have copied text from an old file to correct a DMR network configuration after I messed around with it. The two files that you may have changed in the Expert area are 'mmdvmhost' and 'dmrgateway.'

You can restore the configuration from the Zip backup file if you want to revert to an older configuration or copy your hotspot configuration to a new hotspot.

Apply Changes is used every time you make any changes anywhere in the configuration. It can have unpredictable results. New options can appear on the configuration forms, and it often disables one or more networks by turning off the Enable switches. Especially networks 3 and 4. If this happens after you use **Apply Changes** on the Configuration tab, Set **Enabled back to one** on the Expert Full Edit DMR GW page and click the **Apply Changes** at the bottom of that tab only.

RadioID.net databases

You will probably have already visited the RadioID.net website, to get your DMR ID, and you may have subscribed to the contact list generator. This chapter explains some of the other information that is available at RadioID.net.

You can find someone's name, and DMR ID, by entering their callsign on Radioid.net and clicking **Database** and **DMR User ID Search**. You can also search for repeater information using the **DMR Rptr ID Search** button.

DMR USER ID SEARCH   DMR RPTR ID SEARCH

Repeater Search

RPTR ID	Equals	
Callsign	Begins with	ZL3
City	Equals	
State/Prov	Equals	
Country	Equals	
Frequency	Equals	
Trustee	Equals	

Search

Repeater Results: 5

RPTR ID	CALLSIGN	City	State/Prov	Country	Frequency	Color Code	Time Slots	Offset	Trustee	Network	Talk Groups	Map
530003	ZL3CAR	Christchurch	South Island	New Zealand	438.40000	1	1	-5.00	ZL3TMB	BrandMeister	Details	View
530010	ZL3ADB	Westport 7825	South Island	New Zealand	439.82500	1	Mixed Mode	-5.000	ZL3ADB	BM	Details	View
530301	ZL3DMR	Christchurch	South Island	New Zealand	439.70000	1	TS1 TS2	-5.000	ZL3VP	ZL-TRBO	Details	View
530304	ZL3DMH	Christchurch	South Island	New Zealand	439.98750	1	Mixed Mode	-5.000	ZL3DMH	Brandmeister	Details	View
530305	ZL3DVR	Christchurch	South Island	New Zealand	439.57500	1	Mixed Mode	-5.000	ZL3DMH	Brandmeister + DMR Gateway	Details	View

Figure 45: Repeater search options and results

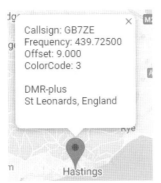

Callsign: GB7ZE
Frequency: 439.72500
Offset: 9.000
ColorCode: 3

DMR-plus
St Leonards, England

Hastings

The Repeater map can be zoomed (CTRL + mouse wheel) to show the location of repeaters in any country. Clicking on a number centres the map on that region. Clicking on an individual repeater teardrop icon shows an information screen which usually includes the callsign, frequencies, DMR network, and colour code. It may also include a contact person and talk group information.

SMS (short message system)

I don't know if many people use the SMS message system, but it's there so I had better cover it. Basically, it is "texting over DMR." The SMS signalling system is also used to carry your APRS beacon, which on DMR is essentially a text message sent to the APRS.fi server.

You do not want to use SMS on Group Call talk groups because the message will go to everybody who is monitoring the talk group. That might be OK if it is "Happy New Year," but not for "Can you please pick up some milk on your way home." SMS should be limited to private ID calls and possibly TG 9 calls on a repeater to advise the time of a Net or some club event.

TIP: It does not matter what talk group the radio is set to, because the text SMS message will be directed at a private call ID through the DMR network. The network forwards the message to the last heard channel on the target radio. That is not necessarily the channel the radio is switched to. If the radio has been switched to a talk group where it can no longer hear that 'last heard' channel, the text message will not be received. This seems to be a bigger problem with the Radioddity radio than it is with the TYT radio.

To complicate matters further, the AnyTone/Radioddity radios can send an SMS using three different data formats. It needs to be set to the Motorola format for APRS or SMS text messaging. **Menu > Settings > Radio Set > Other Func > 24 SMS Format > M-SMS**. You can also compose a message using the keys on the radio or send one of eighteen preset messages. They can be edited on the radio, or the **Digital > Prefabricated SMS** page in the CPS.

The TYT radios can transmit pre-set **Quick Text** messages that are written in the CPS or on the radio. Or you can create messages using the keys on the radio.

SENDING AN SMS MESSAGE

Sending an SMS 'text' message to another radio is easy although I had a lot of problems getting it to work initially. Both radios have to be on the same DMR network or be on the same simplex channel. They do not have to be monitoring the same talk group. The network will send the private call SMS message to the talk group that was "last heard." That should be fine if you are still on that channel, or you set the radio to your monitor channel. If in doubt key up your radio on the channel before you send the text message, just to make sure. SMS is not very reliable. If the other radio cannot hear its 'last heard' channel, the message will not be received.

Radioddity/AnyTone radio

To send a text use **Menu > SMS > New Message >** *Enter your message* **> Confirm > Send**. Or **Menu > SMS > Quick Text >** *Pick and edit a preset message* **> Confirm > Send**.

Then select **Talk Group** and pick a private ID contact (not a group call talk group) or select **Manual Dial** and enter a private ID contact number. (It can be your DMR ID if you have two radios)

TYT RADIO

On the MD-UV380 and similar radios,

Menu > **Messages** > **Write** > *Enter your message* > **Confirm** > **Send**, or
Menu > **Messages** > **Quick Text** > *Pick and edit a preset message* > **Confirm** > **Send**

Then select **Contacts** and pick a private ID contact (not a group call talk group) or select **Manual Dial** and enter the private ID contact number.

The radio display sticks on 'Sending to:'. Change the channel switch to clear it.

SMS WEATHER REPORT

Sending the characters 'wx' to Brandmeister talk group 262993 will respond with a text message containing weather information for your current location.

SMS OVER APRS

You can SMS text to a cell phone over APRS using SMSGTE and the Brandmeister network. I have not been able to make this work, but I expect that is because my local Brandmeister server is in another country. I sometimes get an acknowledgement from the SMSGTE, but I never receive the text on my phone.

Tip: You are supposed to send your message to the local APRS server, 310999 in the USA, 302999 in Canada, 505999 in Australia, 234999 in the UK etc. I believe that the Australian one does not work and that talk group 505999 is assigned to something else. However, if I send the message to 310999 in the USA, the acknowledgement comes back from VK SMSGTE, so I guess that is OK.

There are videos about this topic at
https://www.youtube.com/watch?v=APBSAI8Pjbc&t=1219s
https://www.youtube.com/watch?v=rAq19k6DmIM

Send SMSGTE @63123456789 Hello World on talk group 310999

63123456789 is replaced with your cell phone number. See the note about the correct talk group above.

Hopefully, this will send the text to your phone number. You can reply by sending a text back to the phone number that sent you the message. It needs to include the callsign you are sending to. For example, @YY1ADB-7 *your message*, then hit **send**.

Setting microphone and speaker levels

It is important to set your microphone level so that it is not set so high that it distorts your transmitted audio. I heard one station that was virtually unreadable. Of course, speaking loudly or too close to the microphone can also cause problems. Don't get grumpy if somebody tells you to talk further away from the microphone. They are doing you a favour, pointing out that there is a problem. I am sure that they would not mention it unless the distortion was quite severe. It is possible that clicky audio and dropouts are being caused by network issues, but it is worth checking out. Call a local and ask how you sound.

CHECKING YOUR TRANSMIT AUDIO LEVEL

Brandmeister Hoseline Player signal strength meter

If you have access to the Brandmeister network you can use a feature of the Hoseline Player, to check your microphone level. Open the Brandmeister Hoseline from the Brandmeister dashboard (see page 143) and click Player (top right). Then select talk group 98 on the dropdown list. The white line below the talk groups is a signal meter. Hover the mouse to see the signal level and the peak level in dBm. This is very handy. If you monitor TG 98 in Hoseline Player and make a test call on TG 98, you can monitor your transmission.

Set the radio's microphone level for a signal that just peaks to green on the signal strength meter. You will see the strength that other stations are using. Your signal should never reach the red indicator. It will become distorted at that point.

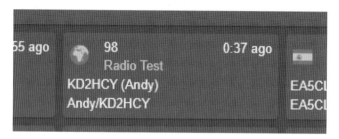

Figure 46: Using Hoseline Player to check transmit audio

The CPS mic gain setting on the AnyTone/Radioddity is at **Common Setting > Optional Setting Volume/Audio > Digi Mic Gain**. On the TYT CPS, it is at **General Setting > Mic Level**.

On the radio, the mic gain setting for the AnyTone/Radioddity is at **Menu > Settings > Radio Set > Voice Func > Digi Mic Level**. On the TYT radio it is at **Menu > Utilities > Program Radio > Mic Level**.

I am using microphone level 4 on the Radioddity and 3 on the TYT.

On other networks

On other networks, you could find a busy talk group where most of the users have about the same audio level, and then, without touching your volume control, switch to the Echo Test talk group and see if your audio level matches theirs.

When an experienced DMR operator tells you your audio gain is too high or too low, listen to them, and make an adjustment.

MINIMUM SPEAKER LEVEL

The AnyTone/Radioddity radios have a range of minimum audio levels labelled maximum volume levels. The audio cuts off quite suddenly when you turn down the volume control, so it is nice to have some control over where that happens. It is very annoying if you turn the volume down and it is still way too loud when suddenly it cuts off completely. I think that is due to the volume control using insufficient bits. The minimum volume can be set in the CPS at **Common Setting > Optional Setting > Volume/Audio > Maximum Volume**. I use 'Indoors' which is the lowest setting. You have to scroll up from the default setting of 1. You could use a higher setting outside where there may be more background noise. On the radio, the adjustment is at **Menu > Settings > Radio Set > Voice Func > Max Volume Level**. There is no equivalent setting on the TYT radio, but the volume seems to be more controllable.

You can set the maximum headphone level as well using **Common Setting > Optional Setting > Volume/Audio > Max Headphone Volume**.

Hotspots

If you have a local DMR repeater and you are happy talking to users of your local DMR network, you do not need a hotspot. However, if you do not have access to a local repeater, or you want the freedom to choose any talk group on any network, a hotspot is the way to achieve that. A word of warning though. Many people "get into DMR" because the radios are cheap. But by the time you add the cost of a hotspot, the DMR option could end up more expensive than buying a D-Star or Fusion radio. Buying hotspots is a bit addictive. The next thing, you will want two so you can add another DMR network, or three, or four so you can monitor multiple talk groups.

Most of the hotspots "out there" are MMDVM simplex hotspots. They transmit and receive on the same frequency, and they only use time slot 2. Most models can be used for some, or all, of these digital voice modes, DMR, D-Star, System Fusion, C4FM, NXDN, POCSAG, and P25. But I am only going to talk about DMR. Most MMDVM hotspots use a Raspberry Pi Zero W or the Pi Zero 2W single-board computer to run the Pi-Star routing software. All of those hotspots can also run on a Raspberry Pi 3, or Pi 3 Model B, but the larger form factor does not match the size of the hotspot. Duplex hotspots are bigger and need more computing power, so a Raspberry Pi 3 or 3B is preferred. The Pi 3 supports 2.4 GHz and 5 GHz WiFi, but the Pi Zero only supports 2.4 GHz WiFi. Raspberry Pi Zero boards are much slower, particularly when booting up, but the gap is narrower with the new Pi Zero 2 which is five times as fast as the old Pi Zero. Some hotspots use other single board computers.

Duplex hotspots have two antennas and they run in duplex mode like a repeater.

Note that the Pi3 setup requires a bigger power supply (2.5A recommended, 700mA minimum) than a Pi Zero and OLED screen setup. I have had no problems running off a powered USB 3 hub. The 2.4" Nextion screen draws 90 mA. You need an additional USB to TTL adapter to use Nextion screens with most simplex hotspots.

MMDVM RASPBERRY PI HOTSPOTS

There are dozens of different models which may come with cases or plastic protection. Many require you to buy a Raspberry Pi Zero or model 3 elsewhere. Here are a few, but a search online will bring up many options. Note I am not endorsing any model. I have not tried most of them.

TIP: It is often difficult to work out what you are going to get. Especially with the Chinese suppliers. For example, some of the websites show a picture of a hotspot with a Pi Zero. But they do not ship with a Pi Zero. The rule is. If the advert does not specifically say something is included… it is probably NOT included. Even if there is a photo showing the item. Some hotspots are supplied assembled, some are not. Some include a case, and some don't.

BI7JTA duplex hotspot. Unfortunately, the model I bought is no longer available. But similar models are available online. It is a duplex hotspot that was provided with a 2.4″ Nextion screen, the programmed SD card and the Raspberry Pi 3 model B. https://www.bi7jta.org/cart/. Apparently, the cost of the STM32 chip has increased dramatically due to short supply.

TGIF Spot. https://tgifspot.com/ has a range of simplex hotspots. One with an OLED display, one with a 2.4″ Nextion display, and one with a 3.5″ Nextion display.

RFinder created the SkyBridge+ dual-band Simplex and the HCP-1 duplex hotspot. The HCP-1 includes an internal battery and is very portable. The computer is a Raspberry Pi Zero.

LZ duplex and LZ simplex hotspots, some come with a 3.2 Inch Color Screen. They are MMDVM hotspots designed for a Rpi Zero

Jumbospot dual-band simplex MMDVM - Rpi Zero or 3B. These are a clone of the ZUMspot. Sold from a wide variety of vendors with a wide variety of prices depending on what you get. Most often they do not include the Raspberry Pi Zero, and sometimes they do not include the case.

ZUMspot https://www.hamradio.com/detail.cfm?pid=H0-016491. These are simplex, UHF hotspots with a 1.3″ OLED screen. They require you to buy a Raspberry Pi Zero or Zero 2. They will also run on an RPi 3 or an Odroid SBC.

Rugged Spot supplies a series of 'NEX-GEN' simplex MMDVM hotspots, some with Nextion screens and some with ceramic antennas (which work very well). They are supplied pre-programmed which is very nice. They are based around a JumboSpot supplied with a Raspberry Pi3-B and a plastic layer case. https://hamradio1.com/product/rugged-spot-store/

TIP: Don't worry if the hotspot is not assembled. All you will have to do is solder the RPi header pins and sometimes the RF SMA connectors (make sure they are supplied though). My simplex hotspot kit came with all the required screws and two short headers for the Raspberry Pi. The hotspot board was 100% complete. You will need an SD card and some free software to create the Pi-Star image and probably a micro USB cable or a USB power supply.

OTHER HOTSPOTS

There are other hotspots. Some have battery power and can be carried around in your pocket.

The SharkRF OpenSpot 3 has a built-in battery. It is a fully cased simplex hotspot with WiFi. The OpenSpot 3 has propriety access software (not Pi-Star). Fast bootup. Can transcode between different digital voice modes. Not cheap and not currently available.

SIMPLEX HOTSPOTS

Most people use simplex hotspots. They are cheaper, easier to set up, and faster to use than duplex hotspots. Simplex hotspots receive and transmit on the same frequency, and they always use time slot 2. Almost all simplex hotspots are MMDVM modem boards paired with Raspberry Pi Zero W or Zero 2 W single-board computers.

The OpenSPOT3 by SharkRF is an exception. It is a self-contained simplex hotspot that can be slipped into your pocket and carried around. The OpenSPOT3 is not an MMDVM hotspot. Instead of Pi-Star, you use 'SharkRF link' for configuration using any web browser. Configuration and connection to a WiFi network seem to be very easy. The only downsides are, the unit can only connect to one DMR network at a time, and they are more expensive than the MMDVM simplex hotspots.

Raspberry Pi zero

You will probably have to buy a Raspberry Pi Zero to supply the computing power for your MMVM hotspot. Most hotspots are supplied without one. You must get the 'Zero W' version which has WiFi. I recommend getting the new Raspberry Pi Zero 2W. It is five times faster than the old Zero W and it still only costs $15 US, (less than £13 UK).

The SD card and display

If you are providing your own Raspberry Pi, you will also have to buy and prepare a micro SD card for the Raspberry Pi. It contains the Linux operating system for the Raspberry Pi and the Pi-Star modem software for the hotspot. I cover that in the 'Loading Pi-Star' chapter starting on page 166.

Because of the small form factor of the MMDVM modem board and the Raspberry Pi Zero, most simplex hotspots have a small 1.3" OLED display or no display at all. But, yes, you can connect a Nextion display to the Raspberry Pi on a simplex hotspot using a USB to TTL adapter.

TIP: The Nextion displays work better with a duplex hotspot on a Raspberry Pi 3B.

Building a simplex hotspot

Don't be scared of building a simplex hotspot. It is just an assembly job. The MMDVM modem board will probably be supplied 100% complete. No soldering is required. You may be required to solder in some header pins on the Raspberry Pi Zero. No problem if you have a temperature-controlled soldering iron and at least some experience with soldering components onto a printed circuit board.

The header pins were supplied in the hotspot kit. I will step you through the process. It is easy. My hotspot kit also included the case, screws, and spacers.

The picture shows the kit that I received. The pcb was supplied in an anti-static bag and no soldering is required on the hotspot board. Just plug the hotspot board onto the Rpi.

Figure 47: A typical simplex hotspot kit. No soldering required.

The case clips together and it is a bit tricky to open. I wedged one side open with a small flat screwdriver and slipped a strip of cardboard into the gap. Then I applied the screwdriver to the other side and the case slid apart easily. Don't reassemble the case until the board has been mated with the RPi Zero and screwed onto the base section. And remember to remove the protective plastic from the screen by pulling the small tab.

I chose to pair the MMDVM modem with the Raspberry Pi Zero 2W. They usually have no header pins installed, but they were supplied in the hotspot kit. If the Raspberry Pi Zero has the full header row you can throw the supplied header pins away.

Solder the header pins into the rows at both ends of the board. Solder the short pins to the PCB. The long side should face up.

Figure 48: Raspberry Pi Zero 2W (no header pins)

I put both sets of pins into the board before I flip the board over so that it sits flat when it is upside down for soldering. The plastic must be flat on the board so that the pins are exactly vertical.

My Pi Zero was supplied with header pins.

Figure 50: Raspberry Pi Zero W (with header pins)

Figure 49: Snip these four leads short if the RPi has a full row of header pins

Use side cutters to trim the four pins on the display board so they won't short out pins on the Raspberry Pi GPIO header. This is especially important if the Raspberry Pi has the full row of header pins.

Place the Raspberry Pi into the case making sure that the SD card slot is accessible. Insert the two short screws on the header side of the Raspberry pi board. Carefully stack the boards, using the supplied spacers on the side furthest from the header plugs. Insert the long screws through the hotspot modem board, the spacers and the Raspberry Pi and tighten, (not too tight).

You might prefer to power up and configure the board before installing the top of the case. Clip the case on being careful not to damage the micro USB and HDMI jacks on the side. Ease the case around them and they will pop into the shell. Screw the antenna onto the SMA connector and power up the board. The micro USB closest to the end is the power connector.

Brandmeister network

Static talk groups can be set up on the Brandmeister dashboard, Hotspot page. See page 145 about how to set up your choice of static talk groups.

If you key up a Brandmeister talk group, it will be activated as **auto-static**. As long as you don't key up a different talk group, the auto-static talk group will not time out and will remain linked until you manually disconnect from it by making a group call to TG 4000 (disconnect), or you key up another talk group. If you key up a second talk group, it will become auto-static, and the previous talk group will become dynamic. The dynamic talk group will time out after 15 minutes or immediately if you manually disconnect both talk groups using TG 4000. TG 4000 does not disconnect static talk groups. Only auto-static and dynamic talk groups.

TAC channels shouldn't be used as primary calling channels. The U.S. TAC channels (310 - 319) cannot be added as static talk groups and they won't become auto-static. If you want a primary calling channel, try one like U.S. Wide 3100.

Dynamic talk groups. Keying up a second talk group makes the new one auto-static and the earlier one dynamic for 15 minutes.

While the previous talk group is dynamic, you'll continue to receive transmissions from it, unless the hold-off timer is active. Keep in mind that if the dynamic talk group is busy, it might overwhelm the new auto-static talk group. If the people using the dynamic talk group aren't leaving adequate gaps between their transmissions, it can be challenging for you to make a call on the new auto-static talk group. If that happens, disconnect both talk groups using TG 4000 and then key up the talk group that you want to listen to.

TIP: Add a TG 4000 (group call) channel to each zone in your radio that contains Brandmeister talk groups, so you can quickly drop a talk group by switching to that channel and pressing the PTT.

TGIF network

The TGIF network operates in the same way as the Brandmeister network, except that TG 4000 is not used. Just key up a channel to make it auto-static. When you key up a second channel it becomes auto-static and the prior one becomes dynamic. You can also set auto-static talk groups on the Self Care page on the TGIF dashboard.

Other networks

The other networks don't have the auto-static mode and you have a more limited opportunity to set up static hotspots. You have to accept the static hotspots that are available on your chosen DMR network server or repeater. If you have a real need such as regular calls to someone in another country, you may be able to contact the repeater or IPSC2 server manager and request that a talk group be made static.

You need to set your Pi-Star dashboard for the talk groups that you want to be static, choosing only talk groups that are static on the network or repeater.

Some will be on TS1, and some will be on TS2, but your simplex hotspot configuration will make them all come out on TS2 as far as your radio is concerned. I cover this configuration in the PiStar chapter.

Keying up a talk group that is not static on the network makes a dynamic talk group connection. It works the same as a static talk group, but it will time out and disappear after 15 minutes (usually) of inactivity.

If you make a group call to TG 4000 (disconnect) you will disconnect the hotspot from the dynamic talk group if there is one active, but not from your nominated static talk groups.

Pi-Star notes for a simplex hotspot

The Pi-Star configuration is covered in the Loading Pi-Star and Pi-Star configuration chapters. There are some minor differences when using a simplex hotspot. Click **Apply Changes** regularly after each configuration step. It takes some time for the modem to reboot, but the configuration screen changes as a result of your previous choices. If something looks odd, such as two entry boxes for frequencies, click Apply Changes and the page will reformat.

- Controller mode is Simplex
- Only one frequency
- Display option is usually OLED /dev/tty/AMA0
- The Radio / Modem type is different

Figure 51: My MMDVM simplex hotspot

DUPLEX HOTSPOTS

More and more people are using duplex hotspots, to take advantage of the two timeslots offered by DMR. There is no advantage in using a duplex hotspot for other digital voice technologies. Duplex hotspots are larger and more expensive. They will work with a Raspberry Pi Zero, but the form factor and computing requirements make a Raspberry Pi 3 Model B a much better option.

Duplex hotspots act like a repeater, transmitting and receiving both timeslots using the DMR data structure. Simplex hotspots use a single frequency and only transmit and receive on TS2. They are a little more complicated to set up, and they are fractionally slower because some time slot setup data has to be sent each time you transmit. The circuit board is typically about twice the width of a simplex hotspot with a 'form factor' designed for mounting on a Raspberry Pi 3 board rather than a RPi Zero. They have two RF chips and two antennas, and they use two frequencies. You can set any frequency split between receive and transmit, but I believe it is best to stick with the standard ±5 MHz offset on the 70cm band and ±600 kHz offset on the 2m band.

Duplex Advantages

I believe that duplex hotspots offer significant advantages over simplex hotspots. If you don't already own a hotspot, I recommend buying a duplex version.

A duplex hotspot can display activity on two selected static or dynamic talk groups at once, routing them to different time slots. You could allocate local talk groups to TS1 and overseas talk groups to TS2, or DMR+ talk groups on TS1 and TGIF talk groups on TS2. If you use the dual slot digital monitor on your AnyTone or Radioddity radio, you will be able to hear calls on either time slot.

You can even transmit on a time slot when your radio is in the middle of receiving a signal on the other time slot. This makes it easier to unlink from a busy talk group even when the hams using the talk group don't leave gaps between their transmissions. A simplex hotspot will not receive your call while it is transmitting.

Time slot allocation

I use TS1 for Brandmeister talk groups and TS2 for DMR+ talk groups. That way if somebody activates one of my selected static talk groups, the hotspot will show it and send it to the radio. I know immediately whether it is a Brandmeister talk group, or a local ZL-TRBO network talk group / DMR+ one. Whether the radio "hears the call" depends on what channel is selected, and if the digital monitor or receive groups are turned on. How you configure your hotspot is up to you, but a duplex hotspot definitely allows more flexibility.

Static talk groups

You can arrange the static talk groups to suit your operation. On the Brandmeister network, the static talk groups are set up on the Brandmeister dashboard, My Hotspot page. It is very easy to change them any time you want to.

Bigger display

My duplex hotspot is driving a Nextion display giving a nice bright colour display of the DMR activity. You can see a VK caller using Brandmeister talk group 505005 on TS1 and a ZL caller using talk group 530 on the DMR+ network on TS2. This MMDVM hotspot is running on a Raspberry Pi Model 3B and using the Pi-Star software.

Figure 52: A duplex hotspot with two time slots busy

Works as a repeater

My hotspot does work as a repeater. I could chat to another radio on TG 9 but considering the flea power of the hotspot compared to the power of your handheld radio, it is probably better to use simplex for radio to radio calls. Besides, I am the only ham in the family, so I have nobody in the house to talk to! Some people connect an amplifier and outside antennas to create a high power hotspot for the local area.

Disadvantages of a duplex hotspot

Your radio may receive a signal on one time slot. When that drops out it might transfer to a call coming in on the other time slot. Then when the original station starts another over, you might miss the rest of the conversation. This is not really a problem with the hotspot so much as the way the radio responds to two time slots being received at the same time. The radio can use either time slot, but not both, at the same time. It can be confusing. There are group call and private call hang time settings in the radio aimed at preventing the radio from switching to the other time slot while you are having a conversation.

POWERING UP YOUR HOTSPOT

Make sure the antenna or antennas are plugged in. Some models have a ceramic antenna on the board, but most have a small SMA antenna, (or two).

Plug the power cable into the Raspberry Pi. It uses a micro USB connector. You can power the Rpi Zero from a plug pack, a USB 3.0 port on your computer, a USB hub, or a USB power supply. A Raspberry Pi 3 is a bit more power-hungry than a Pi zero, especially if you have a Nextion display. Most phone chargers and USB2.0 ports cannot supply enough power and the device may not work, or not work reliably. The official Raspberry Pi power supply is capable of supplying 2.5 amps at 5.1 volts. You may also need a 'micro USB' to 'USB type A' cable.

A Raspberry Pi 3 will boot quite quickly. If you have a Nextion display it will show the idle screen as soon as the power is applied. The screen pages are coded into the display, not the Raspberry Pi. It will take a minute or so for the WiFI to get established with a DHCP network address. If you have a Nextion screen the IP address might be listed at the bottom right of the screen. Eth0 means a wired Ethernet Connection and wlan0 indicates a WiFI connection.

The Raspberry Pi Zero will take a minute to boot, followed by some more time for the WiFI to get established with a DHCP network address.

TIP: If it is the first boot and you copied the wpa_supplicant.conf into the SD card boot drive, the Rpi will boot, load the WiFi config, and then reboot. It could take several minutes.

MY NEW HOTSPOT DOESN'T WORK

Don't panic! I covered this situation in the troubleshooting section, on page 203. The most common problems are,

- An insufficient power supply. Symptoms, include Windows 'bonging' regularly, hotspot rebooting, no display or incorrect display.
- Display not configured correctly in Pi-Star. No display even when rebooting the hotspot.
- Incorrect Modem setting in Pi-Star. No transmission to the radio. No display.
- Hotspot being well off frequency. Seeing activity but no audio from the radio, no display of calling stations, hotspot not responding when you transmit.
- Being on the wrong zone (I was most confused when trying to transmit on my other hotspot's frequency).
- Incorrect colour code, talk group or channel settings.
- Network disabled in Pi-Star. Check the bottom right box on the Pi-Star dashboard. The software has a bad habit of turning networks off, particularly network 3 and network 4.

Loading Pi-Star

Almost all hotspots use the Pi-Star modem software. Its function is to manage routing the data traffic received by the hotspot board to the DMR network. It also routes the traffic from the DMR network to the hotspot board which transmits it to your handheld or mobile radio. On all networks other than Brandmeister and TGIF it manages the static talk groups. If you have a duplex hotspot or repeater, Pi-Star manages which traffic is sent to each time slot.

Pi-Star is extremely capable and if you get into the 'Expert Level' functions very complicated. We will start with a basic setup and proceed from there with caution. I only have a very superficial understanding of Pi-Star, "just enough to get me into trouble." This quote from the website sums it up nicely.

"The design concept is simple. Provide the complex services and configuration for Digital Voice on Amateur Radio in a way that makes it easily accessible to anyone just starting out but make it configurable enough to be interesting for those of us who can't help but tinker." (Pi-Star UK).

Pi-Star runs on the Raspberry Pi. The dashboard interface is just a web page that can be accessed using your favourite Internet web browser. You make changes on the web page and save them back to the hotspot. You can access the Pi-Star dashboard from your PC, a tablet, iPad, or even a phone. Normally for a home hotspot, your controlling device has to be connected to your local LAN at home. You can configure Pi-Star for public access over the Internet and you sometimes see this with publicly accessible repeaters and high power hotspots.

Pi-Star includes the Raspbian Linux distribution from the Raspberry Pi Foundation. It is a variation on Debian Linux that has been optimised to run on the Raspberry Pi ARM boards. Check out the 'What is Pi-Star' page on the Pi-Star.uk website.

THE PI-STAR WEBSITE

The Pi-Star website is at https://www.pistar.uk/. Note that if you have a Pi-Star hotspot plugged in and you type 'Pi-Star' into the URL box, you are likely to end up on your Pi-Star dashboard instead of the Pi-Star website. Pi-Star is based on the DStarRepeater and ircDDBGateway software designed by Jonathan Naylor G4KLX, which has been extended to support the full G4KLX MMDVM suite, including the extra cross-mode gateways added by José (Andy) Uribe, CA6JAU.

The website has a wealth of useful DMR information as well as downloads for the latest revision of Pi-Star for various hardware platforms. There are sections for the Brandmeister, DMR+, and TGIF networks and an AnyTone config page.

SD CARD

Unless your hotspot came with a fully programmed SD card, you will have to buy one. You can fit Pi-Star onto a 2Gb micro SD HC card, but you probably will not be able to find one that small. I ended up with a 32 Gb card and have 27 Gb free space. The Raspberry Pi is not tremendously fast, so there is no need to buy a superfast SD card. The card will probably already be formatted, but if not, you can format it with the FAT32 option. The image will overwrite the formatting with a Linux format anyway. SD cards are available from most electronics and computer shops, or the usual online sources. I have had good results using SanDisk cards, but any micro SD HC card should be fine. Kingston and Samsung disks are also recommended. I don't like Adata products. I had a bad experience with one of their SSDs.

TIP: If the card has already been formatted for Linux, Windows will flip out and try to open each partition. It will also ask you to format each partition. Just keep clicking the close button on all the popup windows until it quits asking. This is a real pain, and it will happen any time you put a Linux SD card into a Windows machine. If you happen to have a Linux machine you can bypass this problem by flashing the SD card on that.

Figure 53: A micro SD HC card and free adapter

SD card reader & writer

If you have a notebook PC, it may have a built-in SD card reader. If it does, it will probably be for the full-size SD cards, but you can buy a micro SD HC card that comes bundled with an adapter. If like me, you use a desktop PC without an SD card reader, you can buy a USB SD card reader (they write as well). They only cost $10 or £4 or thereabouts. You can use one that takes full-size SD cards and buy a micro SD card that is packaged with an adapter, or you can buy one that takes the micro SD card. I have one of each. One of them came free with a unit I bought from China. USB card readers are available from electronics and computer shops, or the usual online sources.

Downloading the Pi-Star image

Unless your hotspot came supplied with a pre-programmed SD card for the Raspberry Pi, the Pi-Star software has to be downloaded from the Pi-Star website.

This is a big 612 Mb download. It will require a good Internet connection and it can take a long time, depending on your broadband speeds. The download took about 4 minutes at my place.

Open the Pi-Star.uk website in your favourite web browser and select **Downloads >**
Download Pi-Star from the menu bar on the left. Or go straight to
https://www.pistar.uk/downloads/. There are download options for several single-
board computers. NanoPi Air, Nano Pi Neo, Odroid XU4, Orange Pi Zero, and the
DV Mega dongle. There may be several releases for the Raspberry Pi listed. The RPi
versions are OK for the Pi Zero and the Pi 3B. Download the newest release by clicking
the orange text. The current release is **Pi-Star_RPi_V4.1.5_30-Oct-2021.zip.**

When the zip has finished downloading, unzip the files to a directory. Perhaps a
Pi-Star directory or a Temp directory. It does not matter so long as you can find the
files again. There are two files, a .img file which is the Pi-Star image and a .md5sum
which is a checksum file used to detect whether the image file is corrupted.

Flashing the SD card

The website has a good set of instruction guides for flashing the image to an SD card,
but I am going to step you through the process anyway. There is also a video at
https://www.youtube.com/watch?v=B5G4gYDdJeQ.

There are a few programs that can write an image file to an SD card. I have tried
'Win32 disk imager' and 'Balena Etcher.' They both work well. I usually use Balener
Etcher for Linux SD disks. If you don't already have it, download one or the other and
install it on your PC.

"Send in the clones." If you have a friend with a working hotspot you might like to
make a clone of their image file rather than download the latest version from Pi-
Star.uk. You can use either program to make a clone of a working SD card onto a new
SD card. Note that this is not the same as making a copy of the files. It is copying the
file structure and formatting as well. Technically it should be possible to format the
card on a Linux machine and then copy the files, but I won't guarantee that won't end
up causing you hours of tinkering.

Insert the SD card into your PC, either directly or in the USB to SD adapter. As
discussed above you may have to use a micro SD to SD adapter as well. These come
free with many micro SD cards. Close any annoying error popup windows from
Windows. Don't follow the advice to format the partitions. **Make 100% sure that you**
know the drive letter of the SD card. You do not want to write a Linux distribution
onto a USB drive you happen to have plugged into your PC.

Balena Etcher method

I like Balena Etcher because it is straightforward and harder to make a mistake. Click
Flash from file and navigate to your .img file. Click **Select Target**. It should find
your SD card automatically but check the drive letter to make sure. Click **Flash**. Wait
until the write and validation cycle is complete. Close the Windows error windows
do not format the disk. Remove the SD disk and the USB adapter from the USB port.

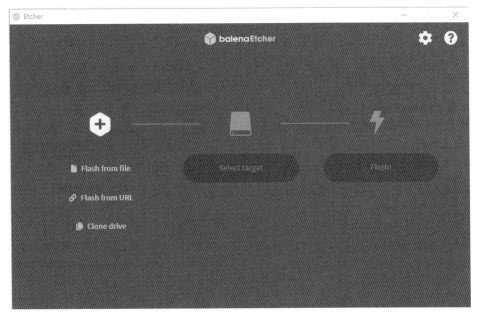

Figure 55: Balena Etcher

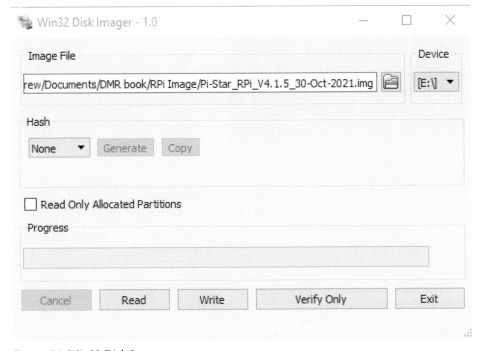

Figure 54: Win32 Disk Imager

Select target 4 found

	Name	Size	Location		
⚠	WDC WD1003FZEX-00MK2A0	1 TB	G:\	Large drive	Source drive
⚠	Samsung PSSD T7 SCSI Disk Device	1 TB	E:\	Large drive	
✓	Mass Storage Device USB Device	32 GB	F:\		

Figure 56: Balena Etcher - choose the SD card

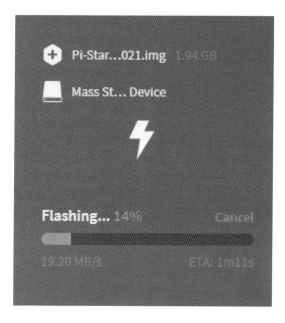

Select the SD card. Note the warnings on the large drives. Check that the 'Size' matches the SD card and note the drive letter.

Click Flash and wait for the disk write and validation to complete.

Cancel any Windows errors and do not format the new disk volumes.

Win32 Disk Imager method

Click the blue disk folder icon to the right of the top 'Image File' text field.

Use the 'Device' dropdown to select the SD card. Check the drive letter to make sure that it is the SD card and not some other drive.

Ignore all the other settings and click the **Write** button. The writing process will be shown on the progress indicator, then the validation process. It can be quite slow.

Wait until the write and verify is complete. Close or Cancel the Windows errors. Do not format the disk. That would overwrite what you have just placed onto the disk. Remove the SD disk and the USB adapter from the USB port.

LAN CONNECTION

Duplex hotspots running on Raspberry Pi 3 or 3B boards can be connected to your LAN with a direct Ethernet connection to your main router, or a hub. I have a four-port internet switch in the shack for connecting my HF transceivers and my PC. However, most people configure the hotspot for WiFi access, and I ended up using a WiFi connection. My duplex hotspot shows the IP address in the bottom right of the idle screen when there has been no talk group activity for a while.

Simplex hotspots always use WiFi because the Raspberry Pi Zero does not have an Ethernet connector.

WIFI CONNECTION

The SD card is complete, but the Raspberry Pi will not have access to your WiFi. Before you proceed you need to know the name of your WiFi network, which is referred to as the SSID (service set identifier) and the WiFi password which is called the PSK (pre-shared key).

TIP: This is the same information that you would use if you were adding a phone or computer to your WiFi network. If you don't know the WiFi name and password, it may be on the back or underside of your WIFI Internet router. Or in the booklet that came with it.

If you are using a Raspberry Pi Zero, you need to complete the steps in configuration option 1 or option 2.

If you are using a Raspberry Pi 3 or 3B, the easiest method is to use configuration option 3. Which involves using the wired Ethernet connection to configure the WiFi connection. But you can use option 1 or option 2 if you like. If you plan to always use a wired Ethernet connection with your Raspberry Pi 3 or 3B, you don't need to set up a WiFi connection at all.

Option 1: Adding the WiFi settings using the Auto AP method

For this method, you need a computer, phone, or tablet with a WiFi connection. It avoids having to download a 'wpa_supplicant.conf' file and copy it into the boot directory of your new SD card. It is also useful if you take your hotspot somewhere and want to connect it to a different WiFi network.

Pi-Star can create its own WiFi network. When you power up the hotspot it will take a while to boot up. The Pi Zero could take up to a minute. If the hotspot is unable to connect to a WiFi access point within two minutes after it finishes booting up, it will create a WiFi access point (AP).

To configure WiFi access so that it can connect to your home network, you disconnect your PC or phone etc. from your usual WiFi network and connect to the Raspberry Pi access point instead. Then you configure the Raspberry Pi for your home network and reboot the hotspot. Lastly, you reconnect your PC, phone etc. to your home network.

1. Boot up the hotspot and wait for two or three minutes until the AP has been created.

2. On a Windows PC, click the WiFi icon on the right side of the toolbar. On a phone, the WiFi settings are in the settings menu. Look for a new WiFi network called Pi-Star-Setup and connect to that. I did not have to enter a password, but if you do, it will be, raspberry.

3. You should be taken directly to the Pi-Star dashboard. If that does not happen, open your web browser and enter pi-star.local into the URL area. After a few seconds, the Configuration page should appear. Or click on Configuration top right.

TIP: Your web browser may want to treat pi-star.local as a search enquiry. If that happens enter http://pi-star.local or http://pi-star instead.

Thanks to W1MSG for his video at
https://www.youtube.com/watch?v=Z5svLP8nEyw

Skip ahead to the WiFi configuration instructions on page 174.

Option 2: Adding the WiFi settings using a manual setup

Pop back to the Pi-Star.uk website and select Pi-Star Tools > WiFi Builder. Enter your country code using the WiFi Country Code dropdown list. In the SSID box enter the name of your WiFi Network. In the PSK box enter your WiFi password, the same as if you were connecting a new phone or laptop to your network at home. Then click Submit.

TIP: If you require a config that will connect to any available open network, leave the SSID and PSK lines empty, the generated config will allow your Pi to connect to any available open network. Then click on Submit.

The website will download a small file called wpa_supplicant.conf.

Open Windows Explorer and look at the boot drive on the SD card. There will be a second drive as well, but you can ignore that. Copy or move the wpa_supplicant.conf file into the SD card boot drive partition.

TIP: The SD Card boot drive should open when you plug the SD card (in its USB adapter) back into the PC. Close or cancel any Windows error popups and do not accept the instruction to format the drive partitions.

Done! Take the SD card out and insert it into the SD card slot on the end of the Raspberry Pi. The card will only go in one way, copper terminations end in first, with the card facing "up" towards the printed circuit board. It will slide in easily. There is no latch.

1. When the card has been inserted into the Raspberry Pi and the Pi-Star system boots up, it will add the config file for the WiFi and then reboot. There may be no indication on the hotspot that this has happened, and it may take a couple of minutes to sort itself out.

2. Open the web browser on your PC and enter pi-star.local into the URL area. After ten seconds, the Configuration page should appear. Or click on **Configuration** top right.

TIP: Your web browser may want to treat pi-star.local as a search enquiry. If that happens enter http://pi-star.local or http://pi-star instead.

Skip ahead to the Pi-Star Configuration chapter.

Option 3: Adding the WiFi settings wired Ethernet method

If you have a Pi3 or 3B you can temporarily connect an Ethernet cable between your LAN router and the hotspot. If you have the situation where the only LAN port is where the fibre or ADSL modem is, and it is remote from the shack, that is not a problem. You can configure the hotspot with the hotspot and the PC separated provided they are still on the same LAN. After the WiFI has been configured you can unplug the hotspot and move it back into the shack.

Open the web browser on your PC and enter pi-star.local into the URL area. After a few seconds, the Configuration page should appear. Or click on **Configuration** top right. If you cannot connect to the hotspot, you will have to find the IP address of the hotspot and enter that. You can find out connected devices by accessing your main router, or by using the FING app. Fing lists it as Pi-Star

TIP: Your web browser may want to treat pi-star.local as a search enquiry. If that happens enter http://pi-star.local or http://pi-star instead.

Continue to the WiFi configuration instructions.

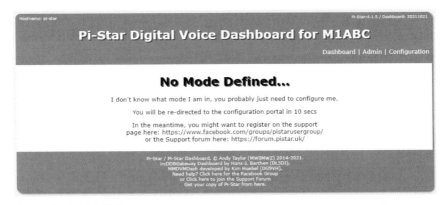

Figure 57: Initial Pi-Star screen

WIFI CONFIGURATION INSTRUCTIONS

If you used option 1 or option 3, you have not finished yet. If you used option 2 the hotspot should already be configured for WiFi and you can skip this section and proceed to the Pi-Star Configuration chapter.

You should have seen the Pi-Star splash screen on the previous page and after 10 seconds the Configuration page should have opened. If not, click Configuration at the top left of the Pi-Star screen. I will cover the general Pi-Star configuration in the next chapter. This section is just for completing the WiFi setup.

Scroll down the Config screen until you reach the 'Wireless Configuration' section. It should look like this. Note that it says that the WiFi interface is down. If it says the WiFi interface is up, the WiFi is already configured and working.

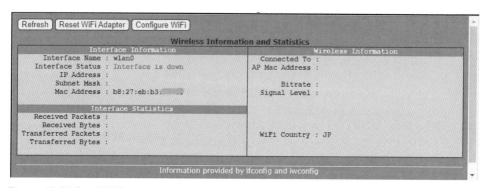

Figure 59: Pi-Star WiFi status

Figure 59: Pi-star WiFi Config.

Just below the status box, there should be a configuration box, like this one. Select your country using the **WiFi Regulatory Domain (Country Code)** dropdown list. I selected NZ. Then press the **Scan for networks (10 seconds)** button. It should provide you with a list of WiFi sources that the hotspot can see. Your home LAN should be near the top because it should have the strongest signal. So, select your WiFi LAN. Alternatively, you can set this up manually by pressing Add Network.

This should populate the Wireless Config. With your Country Code and SSID (WiFi network name). Enter your WiFi password into the PSK box and press Enter.

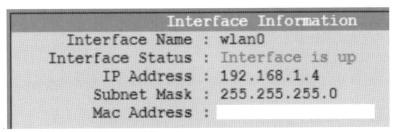

Figure 60: Choosing the WiFi network and entering your WiFi password

Above the 'Wireless Communication' box, there is a button marked **Apply Changes**. Click that. After about 30 seconds the Config screen will go back to the splash screen and then after another 30 seconds or so, it will return to the config screen. You should be rewarded with the green **Interface is up** message.

Take a note of the IP Address just in case you need it later.

```
              Interface Information
    Interface Name : wlan0
   Interface Status : Interface is up
        IP Address : 192.168.1.4
       Subnet Mask : 255.255.255.0
       Mac Address :
```

Figure 61: The WiFi is working :-)

IMPORTANT NOTE

Now you need to restart the hotspot to transfer from the AP mode (option 1) or the Ethernet mode (option 3). At the top of the page, there is **Power > Reboot > OK**. Wait 90 seconds for the reboot to finish, and the dashboard should reset. Or just power down the hotspot and restart it.

If you used option 1 now is the time to reconnect your phone/PC/tablet to your normal network. If you used the Ethernet cable, unplug it now.

The hotspot will jump to a different IP address when it restarts. It should be the one you recorded. That means that your Pi-Star web page may not work anymore. Retry the pi-star.local or http://pi-star.local page several times and it should eventually restart. If that does not work, try entering the IP address as a URL.

Pi-Star Configuration

OK, the hotspot is running, and you have configured the WiFI access, which is "the tricky part." Now Pi-Star has to be configured.

Pi-Star runs on the Raspberry Pi, but you configure it using a webpage interface on your PC, tablet, or phone. This is normally only accessible from within your home network. You can choose to make it public, but that would only be for a public repeater or high powered hotspot.

If you are not already there, with the hotspot booted up and running and Pi-Star on your web browser, plug in the hotspot, start your PC web browser, and type

http://pi-star/ or

http://pi-star.local/ or

http://pi-star/admin/configure.php or

the IP address, usually something like 192.168.1.4

This should bring up the Pi-Star dashboard.

PI-STAR ERROR MESSAGE

You may get an error message like this one. It just means that the modem selected in Pi-Star does not match your hotspot hardware, which is not surprising since you have not set that yet. Just click **OK** and carry on.

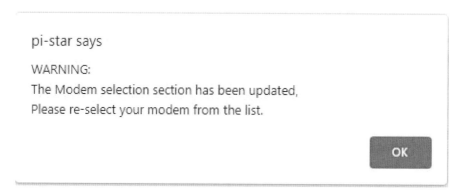

```
pi-star says

WARNING:
The Modem selection section has been updated,
Please re-select your modem from the list.

                                              OK
```

PI-STAR BASIC CONFIGURATION

If you are on the dashboard page, click **Configure** to get started. If you are asked for a login name it is **pi-star** and the password is **raspberry**.

WOW! This looks complicated. Luckily you only need to do a few things.

We are going to start with a basic connection and work up from there.

The top section is 'Gateway Hardware Information.' It contains the hostname, software kernel (revision), the platform (model of Raspberry Pi), CPU loading and the CPU temperature. Green is good. Orange is OK. If it gets into the red, you need a heatsink or a fan on the CPU in your Raspberry Pi.

The next section is 'Control Software.' It should already have **MMDVMHost** selected because we are not configuring a D-Star repeater. If you have a simplex hotspot (one antenna) select **Simplex Node**. If you have a duplex hotspot (two antennas) select **Duplex Repeater**. This is **an important setting**.

The third section, 'MMDVM Host Configuration,' is where we set up for different digital voice modes and for transcoding between voice modes. We are only interested in DMR at present, so click the switch icon to turn **DMR Mode** on. The switch should change to an orange colour. We will leave the **RF Hangtime** and **Net Hangtime** set to the default 20 seconds. You can change them later if you want to. The last line in this section selects the display type if your hotspot has a display.

MMDVDM Display Type

If you have no display screen, set the MMDVDM Display type to **None**.

For an OLED (small display) select OLED type 3 for the very small 0.96' screen or **OLED type 6** for the more common 1.3" screen. Either option works on my hotspot.

If you have a Nextion screen, select **Nextion**.

There are a couple of other options, including one for TFT displays.

Port

For OLED displays, select **/dev/tty/AMA0** because the display is being driven directly by the modem board.

For Nextion displays on you usually select **modem** unless the display is connected to a TTL to USB adapter plugged into the Rpi. In that case select **/dev/tty/USB0**.

Nextion Layout

This selects from four display layout options. Note that the background image is stored on the display itself. This dropdown only changes the information that is sent from Pi-Star to the display. It does not change the basic layout.

The Nextion layouts can be edited on your PC and loaded directly into the Nextion display. You may have seen a power adapter in the Nextion box. See the information at. https://on7lds.net/42/nextion-displays.

Click the **Apply Changes** button to load your changes to the hotspot.

Setting the Nextion screen resulted in my display starting to show the IP address.

The 'General Configuration' section includes some very **important settings**.

Identity

Hostname: **pi-star** I don't see any good reason to change this.

Node Callsign: This is usually your **callsign**. If it was a public repeater, it would be the repeater callsign.

CCS7 / DMR ID: Enter your (or the repeater's) **DMR ID number**.

Radio Frequencies

Radio Frequency: If you selected 'Simplex Node' there will only be one frequency box. If you entered 'Duplex Repeater' and remembered to Apply Changes, there will be a receive frequency and a transmit frequency.

You can use any frequencies that are in the band(s) supported by both the hotspot and your handheld radio. Almost all DMR hotspots operate in the 70cm amateur band. Although the hotspot transmits very low power and is unlikely the interfere with others, you must remember that your handheld may be transmitting a much higher power. Check your local 70cm band plan. There will be band segments for repeaters and for digital simplex. For my simplex hotspot, I chose one of the designated digital simplex frequencies. I believe that it is a good idea to use the standard 5 MHz repeater offset and band segment for a duplex hotspot or repeater on the 70cm band. I selected a repeater pair that is not in use for any other repeaters in New Zealand. Just choose a repeater that is not being used in your region. If somebody complains it is a trivial matter to change to a different frequency, although it will mean changing all the channels on your radio.

Setting		
Hostname:	pi-star	Do not add suffixes
Node Callsign:	ZL3DW	
CCS7/DMR ID:	1234567	
Radio Frequency RX:	438.125.000	MHz
Radio Frequency TX:	433.125.000	MHz

Figure 62: Pi-Star ID and frequencies

Remember that the duplex hotspot frequencies are the reverse of the frequencies that you program into your radio. The radio receives the hotspot transmit (output) frequency and it transmits on the hotspot's receive (input) frequency. I know that you already know this, but it is easy to get it wrong 😊.

Location

Add your **latitude and longitude,** using the degrees and decimal 172.1234 notation rather than degrees minutes and seconds. If you don't know your latitude and longitude, find your street on Google maps click on the map, then right-click, and your location will be displayed. The latitude is a positive number in the northern hemisphere and a negative number in the southern hemisphere. The longitude is measured in degrees East (+ve) or degrees West (-ve) of the zero line.

Enter your closest town or your locality and your Maidenhead grid. For example, Christchurch, RE66hm. Enter your **Country.**

URL you can enter your personal website or select **Auto** to use your QRZ.com listing. Most folks do that. Note that selecting Auto won't change the URL text box until you go through the **Apply Changes** process.

Latitude:	-43.497	degrees (positive value for North, negative for South)
Longitude:	172.605	degrees (positive value for East, negative for West)
Town:	Christchurch, RE66hm	
Country:	New Zealand	
URL:	https://www.qrz.com/db/ZL3DW	◉ Auto ⚪ Manual
Radio/Modem Type:	MMDVM_HS_Hat_Dual Hat (VR2VYE) for Pi (GPIO) ⌄	
Node Type:	◉ Private ⚪ Public	
APRS Host Enable:		
APRS Host:	aunz.aprs2.net ⌄	
System Time Zone:	Pacific/Auckland ⌄	
Dashboard Language:	english_uk ⌄	

Radio modem type

The radio modem type has to suit the hotspot design. You might have to experiment to find the correct one if the manufacturer didn't supply the modem type information. It will be an RPi one, not a Nano or USB stick. Mine is a 'dual hat' (duplex) BI7JT hotspot from China. The firmware was written by VR2VYE. The STM32-DVM / MMDVM_HS – Raspberry Pi Hat (GPIO) modem seems to work with most generic Chinese simplex hotspots. If you bought a ZUMspot try one of those options, if it is a DV-Mega try those options.

Node type

Set the node type to Private. It would only be set to public if your repeater or hotspot was also available publicly. If you do set it to 'public,' change the default password.

APRS Host

If you turn APRS Host Enable on, set the closest APRS host, or rotate.aprs2.net which guarantees a connection to a Tier2 APRS server.

It lets the hotspot announce its position on APRS.fi, rather than the radio. I leave it turned off since my hotpot is not travelling anywhere.

System time zone and language

Set these to your time zone and preferred language option.

DMR Configuration

The 'DMR Configuration' section sets the DMR network settings. This is where you have to make some serious decisions.

DMR Master

If you will only be using one DMR network, select it using the dropdown list. If you plan to set up the hotspot for Brandmeister and a second DMR network, select **DMR Gateway** and **Apply Changes**.

There are 43 **Brandmeister servers** listed. If you wish to use the Brandmeister network you can select any BM server, usually the one closest to your location.

There are 80 **DMR+ servers** listed. If you wish to use the DMR+ network, you should select the DMR+ IPSC2 server closest to your location. Some countries have several options. For example, the USA has three regional options 3102, 3103 and 3103 and two Quadnet options. Some US states have separate IPSC2 servers and there are several HAMNET servers. Spain has four options. New Zealand and a few others have a hotspot only option and a general (hotspots and repeaters) option. Australia has Australia, Australia DMR, and VKHOTSPOT.

If you are subscribed to DV Scotland, Phoenix K, Phoenix E, or Phoenix F there are DMR+ IPSC2 servers for that. Below the DMR+ options, there are servers for the FreeDMR, TGIF, FD, and HB networks. The list also contains DMR2YSF and DMR2NXDN options. They are for using a DMR radio on YSF or NXDN networks. These options are not covered in this book.

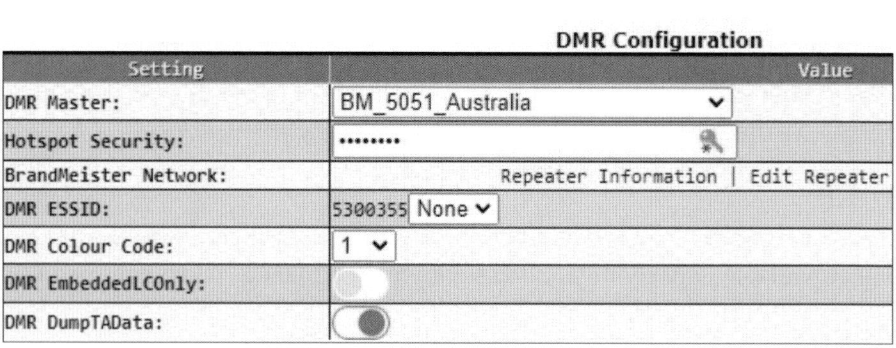

DMR Configuration

Setting	Value	
DMR Master:	BM_5051_Australia ⌄	
Hotspot Security:	•••••••• 🔑	
BrandMeister Network:	Repeater Information	Edit Repeater
DMR ESSID:	5300355 None ⌄	
DMR Colour Code:	1 ⌄	
DMR EmbeddedLCOnly:	◯	
DMR DumpTAData:	◯	

Figure 63: Typical Brandmeister setup

Hotspot Security

If you are using the Brandmeister network or other 'managed networks' such as TGIF, Phoenix, DV Scotland etc. you will have to enter the password for that service.

On Brandmeister and TGIF the password is provided on the Self Care page. Other networks such as DMR+ do not require a password, so don't add one here.

Brandmeister Network

If you selected a Brandmeister server, this line has links to your 'Repeater Information' and 'Edit Repeater' pages on the Brandmeister Self Care web page.

DMR+ and other networks

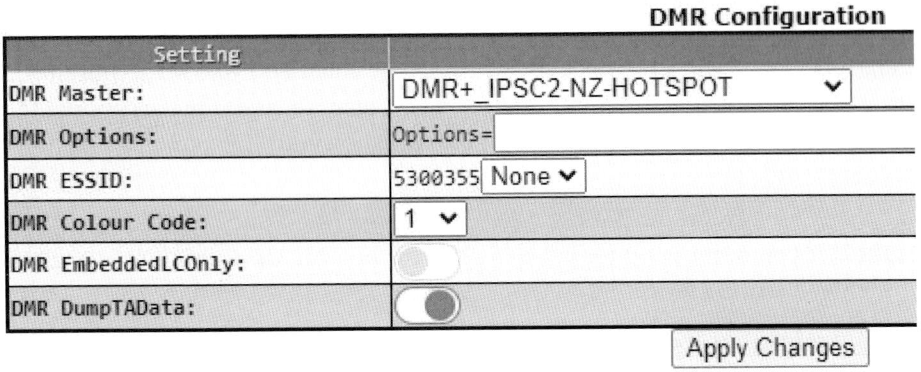

DMR Configuration

Setting	
DMR Master:	DMR+_IPSC2-NZ-HOTSPOT ⌄
DMR Options:	Options=
DMR ESSID:	5300355 None ⌄
DMR Colour Code:	1 ⌄
DMR EmbeddedLCOnly:	
DMR DumpTAData:	

Apply Changes

Figure 64: Typical DMR+ setup

Set **DMR Master** to a DMR+ server near you, or DMR+ Phoenix, TGIF etc.

DMR options is where the reflector and/or static talk groups will be added.

Enter a text string with semi-colon delimiters and no spaces.

UserLink=1; allows the user (that's you) to select a different reflector. UserLink=0 inhibits that function.

StartRef=4xxx; sets a link to reflector 4xxx when you restart the hotspot. This can be TG 4000 in which case the hotspot will send a disconnect when you restart the hotspot.

RelinkTime=15; re-sends the reflector link command every xx minutes. If you connect to a different reflector, this sets the idle time before it will fall back to the reflector mentioned in the StartRef command.

TS1_1 to TS1_5. If you have a duplex hotspot or a repeater, you can set up to five static talk groups on time slot 1. These can only be selected from the choices made available by the server owner. i.e. the talk groups listed as TS1 on the server's IPSC2 dashboard. Don't enter more than you need. The format is TS1_1=235;TS1_2=1;TS1_3=530;

TS2_1 to TS2_5. For simplex or duplex hotspots and repeaters. You can set up to five static talk groups on time slot 2. These can only be selected from the choices made available by the server owner. i.e. the talk groups listed as TS2 on the server's IPSC2 dashboard. Don't enter more than you need. The format is TS2_1=235;TS2_2=1;

Example 1: A complete string might look like

StartRef=4850;RelinkTime=15;UserLink=1;TS1_1=91;TS1_2=1;TS1_3=530;TS2_1=110;TS2_2=270;

What does this do? It links reflector 4850 as a static talk group. If you connect to a different reflector, it will drop back to 4850 after the channel is idle for 15 minutes. It says that you can link to a different reflector if you want to, and it makes several talk groups static so they will automatically be broadcast by your hotspot.

--//--

Example 2: If you don't want to link to a reflector but you do want to set static talk groups, enter a string like this

UserLink=1;TS1_1=91;TS1_2=1;TS1_3=530;TS2_1=110;TS2_2=270;

What does this do? It makes several talk groups static so they will automatically be broadcast by your hotspot.

--//--

Example 3: If you don't want to link to any talk group when the hotspot boots up, set the Options to, StartRef=4000;RelinkTime=60;UserLink=1;TS1_1=9;

What does this do? It sends a disconnect via talk group 4000, every 60 minutes. It says that you can link to a reflector if you want to, and it makes local calls static.

--//--

DMR ESSID

The ESSID is an extension to your DMR ID (or the repeater's DMR ID) to allow you to use two hotspots on the same DMR network. Leave it set to **None** unless you are setting up a second hotspot using your DMR ID, on the same DMR network.

DMR Colour Code

This is the colour code for your hotspot. Most people leave it at **1** unless there is a risk of getting traffic from another repeater or hotspot on the same frequency. Since you will not configure a second hotspot on the same frequency, I recommend leaving it set to 1.

DMR EmbeddedLCOnly. Turning this on disables the Talker Alias feature. Leave it off unless you are experiencing audio dropout problems.

DMR DumpTAData sets the logging of Talker Alias data. The Default is to leave it on, but you can turn it off if you are not using Talker Alias. I left it on.

Click **Apply Changes** to save your settings

Brandmeister Master

If you selected **DMR Gateway** and **Apply Changes**, an additional line will appear for you to enter your preferred Brandmeister server. You usually select the Brandmeister server closest to your location

Brandmeister password problems

Check 'Network Status' on the main dashboard page. It must be green.

I had a lot of trouble getting the Brandmeister password to stick. Every time I saved using Apply Changes, I used to have to enter the password. Otherwise, the DMR NET connection failed. I had to select Brandmeister as the single server and then Apply Changes, then go back and set the DMR Master to DMR Gateway and Brandmeister Master to the BM server. Now the password seems to stay put.

Another issue is the Network Enable switches turning themself off when you Apply Changes. You have to turn them back on and Apply Changes again before the network will work.

MOBILE GPS

If you have a GPS receiver attached to the hotspot, (not the one in the radio), you can enable it, set the device port on the Raspberry Pi (Linux) and set its communication speed. I don't know what GPS data standard is required and I have not experimented with this option.

FIREWALL CONFIGURATION

You can leave everything set to the default options. You can always access the RPi via SSH within your local LAN.

If you need access over the Internet, you can change any or all of the top three settings to Public. These settings are used for accessing the dashboard remotely, from outside your network. To quote Andy Taylor in the Pi-Star Users Support Group: "These settings tell the uPNP daemon to request port forwards from your router."

- Dashboard Access: requests TCP/80

- ircDDBRemote Access: requests UDP/10022

- SSH Access: requests TCP/22

Auto AP enables the automatic AP when the hotspot cannot find a wireless network within two minutes. I suggest you leave it turned **on**.

UPnP (universal plug-n-play) should be set to **on**. It is needed for D-Star port forwarding and SSH access.

Wireless (WiFi) Configuration has already been covered.

Auto AP SSID

This setting changes the default so that a password and login name is required to access the hotspot in AP mode. If you need the AP mode, you are probably in enough trouble without adding this extra layer of complexity.

Remote access password

Use this if you have given the hotspot public access over the Internet. The public will, if they know the IP address or URL, be able to see the Pi-Star dashboard. This password stops them from re-configuring the hotspot.

PI-STAR BACKUP

When you have your settings right, or nearly right, make a backup of the Pi-Star configuration. It will be saved to your downloads area as a Zip file. Click **Configuration > Backup/Restore > Big down arrow.**

You can restore a previous arrangement by selecting the file with **Choose File** and clicking the **Big up arrow**. This could be useful if you make configuration changes that didn't work well, or if you want to save multiple arrangements. For example, you might have a setup that has two DMR networks and another that has two different DMR networks. You could also use the zip file to load a Pi-Star configuration from another hotspot.

UPDATING PI-STAR SOFTWARE AND MMDVM FIRMWARE

WARNING: There is no need to update the Pi-Star software if you just downloaded it from the Pi-Star.uk website. It will already be the latest version. There will probably be no need to update the MMDVM hotspot firmware either unless it is a very old modem. Don't update either unless you have a good reason to do it.

Pi-Star software update

You can update Pi-Star from the Pi-Star dashboard. The version you are running is displayed at the top right of the dashboard page. You probably never noticed it. My dashboard has 'PI-STAR: 4.1.5 / Dashboard: 20211111. This means that the Pi-Star version is 4.1.5 and the Dashboard version was released on the 11th of November 2021.

To update the software, click **Configuration > Update**. A load of green text will show the Linux update happening on the Raspberry Pi. When it has finished, re-boot the hotspot, and the dashboard page will show the new version number if there was an update available.

SSH access to Raspberry Pi

You can access the Raspberry Pi that is running your hotspot with SSH (Secure Shell). Most people use PuTTY for SSH access, but any SSH client will do.

I can't think of any reason to access the Raspberry Pi, but you can do a software update that way if you want to.

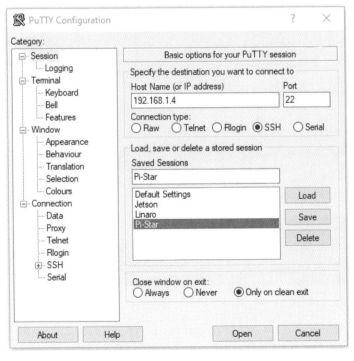

Enter the Pi-Star's **IP address** or pi-star.local. into the host name box. Save it see the note below. Then click **Open**.

I found that using the host name crashed PuTTY but the page would come up after a delay of a minute or more. It is probably something to do with my PC setup.

Using the IP address worked perfectly.

It is a good idea to enter the hostname or IP address and then save it by entering a name in 'Saved Sessions' and clicking the Save button. If you don't do that, this rather ominous warning message will pop up when you click Open. Just ignore it and click **Yes**.

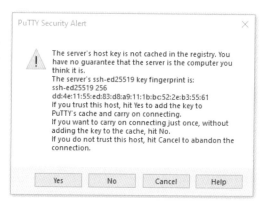

You will be presented with a PuTTY terminal window.

login as: **pi-star**

pi-star@pi-star.local.'s password: **raspberry**

Figure 65: SSH access to the Raspberry Pi

This screen indicates that you have access to the Raspberry Pi. You can use the normal Linux command lines, ls -a to list a directory, df -h to check the SD card space, etc.

Note the instructions on the boot screen that say 'Pi-Star's disk is read-only by default. You can change it to read-write access, at your own risk, using **rpi-rw**. You do not need to do this to run the update or upgrade commands. Also, all pi-Star tools start with a **pistar-** prefix. (Note there is no hyphen between pi and star).

TIP: You can also access the Raspberry Pi by plugging in a USB keyboard and an HDMI monitor, then rebooting the hotspot.

Updating the Pi-Star software from the command line

There is no need to update the Pi-Star software from the command line because you can do it from the Pi-Star dashboard. But you can, using **sudo pistar-update**. I believe that it runs the same command as doing it from the dashboard, but it also updates the Raspbian Linux and other software on the Raspberry Pi. It can take a long time to complete. The update will stop for up to a minute on some lines. Just be patient, go and make a coffee, and let it run until you get the Linux prompt back.

You can also run **sudo pistar-upgrade** which will look for an upgraded version of Pi-Star. Well, that was interesting! When I ran the upgrade command, it said I was already running the latest version (as expected) but it now the dashboard reports that I am running version 4.1.6.

Updating the MMDVM firmware from the command line

The MMDVM firmware is indicated in the Radio Info box on your Pi-Star dashboard. It is updated using a command like **sudo pistar-mmdvmhshatflash hs_hat** or if you own a ZumSpot **sudo pistar-zumspotflash rpi**. You **MUST** know the correct firmware for your hotspot. It should be on the manufacturer's website. **If in any doubt don't do it**.

ADDING OTHER DMR NETWORKS

You can configure Pi-Star to access up to four DMR networks. This is achieved using the Brandmeister 'Gateway,' so Brandmeister must become the first network connection and you must register with Brandmeister. The basic configuration makes the second network a DMR+ IPSC2 server connection, but it is up to you whether you use it. You can leave it disabled, or possibly reconfigure it in the Expert section. I will show you how to configure your MMDVDM hotspot for Brandmeister and DMR+, and then how to add the TGIF network. That should give you enough understanding of the methodology to use a different third network if you wish.

There is an immediate problem with having access to two or three networks on the same hotspot. When a call is made from your radio, how does the hotspot know which network to send it to? This is managed by shifting a block of talk group numbers to a different range. For example, to use TGIF talk group 3118 we make a talk group 5003118, create a channel on our hotspot frequency in the usual way and add the channel to a zone, just like normal. The hotspot subtracts 5000000 from the TG number and sends the call to the TGIF network as TG 3118. A call originating on talk group 3118 on the TGIF network gets 5000000 added before the call is transmitted from the hotspot. We do the same thing with the DMR+ network except the offset is 8000000. I have never tried to connect four networks because the TG commands get complicated.

ADDING THE DMR+ NETWORK

WARNING: To add a second DMR network you need to get into the 'Expert' area of Pi-Star and edit the config files. I am not a Pi-Star expert, so I will explain the settings that work for me. There may be other methods that work just as well, or better. I have not attempted to learn the hundreds of other settings, so play with them at your own risk.

You MUST have a working configuration before you attempt to add a second DMR network! Things can only get worse if you do not.

The very first thing you should do is use **Configuration > Backup/Restore > Big down arrow** to make a backup so that you can get back to a working configuration if everything goes "pear-shaped."

As far as I know, the primary network has to be Brandmeister because you are using the Brandmeister gateway to the other networks. You can add up to three other networks, but the more you add the more complicated it will become. Another way to achieve multiple networks is to have multiple hotspots on different frequencies.

Here are the steps to add a second network to Pi-Star.

1. If you are currently using DMR+ or any network other than Brandmeister, take a note of the current DMR Master and DMR Options settings. Assuming this will become your number two or number three network, you will need the information later. If you are currently using a Brandmeister server, take a note of which one.

2. Change the DMR master setting to **DMR Gateway** and **Apply Changes**.

Setting	Value
DMR Master:	DMRGateway
BrandMeister Master:	BM_5051_Australia
BM Hotspot Security:	●●●●●●●●
BrandMeister Network ESSID:	5300355 None
BrandMeister Network Enable:	●
BrandMeister Network:	Repeater Information | Edit Repeater (BrandMeister Selfcare)
DMR+ Master:	DMR+_IPSC2-Australia
DMR+ Network:	Options= StartRef=4850;RelinkTime=15;UserLink=1;TS1_2=1;TS1_3=530;TS1_4=110;TS1
DMR+ Network ESSID:	5300355 None
DMR+ Network Enable:	●

Figure 66: Two DMR networks. Brandmeister and DMR+

3. After the hotspot restarts, there should be an additional line for you to enter your preferred Brandmeister server. You usually select the Brandmeister server closest to your location. Or the one you were using before.

4. BM Security: enter the Brandmeister password from the Self Care page on the Brandmeister dashboard. *BUG: If the password won't stick see page 183.*

5. The DMR Configuration section will have grown to include four lines for adding the DMR+ network.

6. If you were already using DMR+ or a connected system like Phoenix or ZL-TRBO, you should enter the DMR+ server and DMR Network options that you had before. If not, enter the DMR+ server you want to use and the DMR+ Network Options as discussed above, (page 181).

7. Do not add an ESSID extension unless this is your second hotspot on that network.

8. Turn on Brandmeister Network Enable and DMR+ Network Enable and click **Apply Changes**.

9. On my system turning the DMR+ network on turned the TGIF network off. I had to go and enable it again in the Expert section. Then click the **Apply Changes** button in the Expert section, not the Configuration page.

10. Your Pi-Star dashboard page (bottom left) should show the BM server and the DMR+ server if you enabled it.

11. Are we done? Sadly not!

Talk group offsets

At this stage, the two networks will be sending static talk groups to the hotspot, but the talk group numbers conflict. If we transmit on talk group one, which network should the hotspot send it to? We need to make some 'Expert' changes to shift the DMR+ talk group range. I guess you could shift the Brandmeister talk group range if you prefer.

We are going to use an offset of 8000000. That means to use DMR+ talk group 530 we make a talk group 8000530, create a channel on our hotspot frequency in the usual way and add the channel to a zone, just like normal. The hotspot subtracts 8000000 from the TG number and sends the call to the DMR+ network as TG 530. A call originating on talk group 530 on the DMR+ network gets 8000000 added before the call is transmitted from the hotspot.

Adding the DMR+ network continued

Click **Configuration > Expert > Full Edit: DMR GW**. This opens the screen that writes one of the Pi-Star initialisation files.

Scroll down the list until you see [DMR Network 1]. There is a list of initialisation statements that have resulted from the settings we changed so far. Let's analyse the meaning of the key statements.

TGRewrite

Converts a talk group number and time slot, to a different talk group number and time slot. [from **TS** and **TG** to **TS** and **TG, range**]. RF to network and network to RF.

TGRewrite0=2,8,2,9,1
Converts TS2 TG 8 to TS2 TG 9 with a range of 1. This is used when you have a reflector on network two (DMR+) which requires you to make a 'local TG 9 call.' You can't use TG 9 because that is being used for local calls on the BM network.

TGRewrite1=1,8000001,1,1,999999
Converts TS1 TG 8000001 to TS1 TG 1 on the DMR+ network. The range covers the next 999999 talk group numbers, so talk group 8000530 will be sent to the DMR+ network as 530. 8003120 will be sent as 3120.

Each TGRewrite line gets a sequential number TGRewrite0, TGRewrite1, etc.

PCRewrite

PCRewrite does the same thing, except it works on private Call IDs. RF to network.

PCRewrite0=2,8004000,2,4000,1001
Converts private call TS2 TG 8004000 to TS2 4000 (Disconnect), plus the next 1001 talk groups which should include all the 4000 series reflectors.

PCRewrite1=2,8009990,2,9990,1
Converts TS2 TG 8009990 to TS2 TG 9990, which is Echo on the DMR+ network.

TypeRewrite

TypeRewrite is used to convert a group call into a private call. It only works in the radio to network direction. It also translates the number range like a TGRewrite. This line will stop you from making a group call to Parrot or Echo if the talk group is incorrectly set on the radio

TypeRewrite1=1,8009990,1,9990
Converts TS1 TG 8009990 to TS1 TG 9990 and converts the group call to a private call.

SrcRewrite

SrcRewrite is used to send a group of time slots to a single time slot. It only works in the network to radio direction. It is used to divert all reflectors onto TG 9 on network one (BM), or TG 8 on network 2 (DMR+)

SrcRewrite0=2,4000,2,9,1001
Converts TS2 TG 4000-TG 5001 to TS2 TG 9.

PassAllPC, PassAllTG

These are only used on DMR network 1. They catch anything that is missed by the other rules.

PassAllPC0=1   *pass all Private Call traffic on TS1 irrespective of the TG number*

PassAllTG0=1   *pass all Talk Group traffic on TS1 irrespective of the TG number*

PassAllPC1=2   *pass all Private Call traffic on TS2 irrespective of the TG number*

PassAllTG1=2   *pass all Talk Group traffic on TS2 irrespective of the TG number*

A real example – network 1 (Brandmeister)

[DMR Network 1]   *The square brackets indicate that this is a section header*

Enabled=1   *Enabled = 1 means the function is turned on, 0 = off*

Address=5051.master.brandmeister.network   *This can be an IP address or a URL*

Port=62031   *Port is allocated by the config software*

TypeRewrite0=1,9990,1,9990   *Parrot TS1 to private call*

TypeRewrite1=2,9990,2,9990   *Parrot TS2 to private call*

SrcRewrite0=1,4000,1,9,1000   *Reflector listen on TG 9*

SrcRewrite1=2,4000,2,9,1000   *Reflector listen on TG 9*

PassAllPC0=1   *Pass all Private Calls on TS1*

PassAllTG0=1   *Pass all Talk Groups on TS1*

PassAllPC1=2   *Pass all Private Calls on TS2*

PassAllTG1=2   *Pass all Talk Groups on TS2*

Password="*myBMpassword*"   *Enter your **Brandmeister password***

Debug=0   *Allocated by the config software*

Location=1   *Allocated by the config software*

Id=5300xxx   *Your **DMR ID** allocated on the config page*

Name=BM_5051_Australia   *The BM Server you chose in the config software*

A real example – network 2 (DMR+)

[DMR Network 2]   *Network 2 section header*

Enabled=1   *DMR+ is enabled*

Address=139.180.162.104   *IP address assigned by Config page*

Port=55555   *Port assigned by Config page*

TGRewrite0=2,8,2,9,1	*Changes TS2 TG 8 to TS2 TG 9 for reflectors*
TGRewrite1=2,8000001,1,1,503	*8000001 - 8000504 TS2 to 1 - 504 TS1*
TGRewrite2=2,8000505,2,505,45	*8000505 - 8000550 TS2 to 505 - 550 TS2*
TGRewrite3=2,8000551,1,551,262448	*8000551 - 8262999 TS2 to 551 -262999 TS1*
PCRewrite0=1,8004000,1,4000,1000	*Disconnect and reflectors TS1*
PCRewrite1=2,8004000,2,4000,1000	*Disconnect and reflectors TS2*
PCRewrite2=1,8009990,1,9990,1	*Parrot TS1 TG 8009990 to TG 9990 TS1*
PCRewrite3=2,8009990,2,9990,1	*Parrot TS2 TG 8009990 to TG 9990 TS2*
TypeRewrite1=1,8009990,1,9990	*Parrot TS1 to Private*
TypeRewrite2=2,8009990,2,9990	*Parrot TS2 to Private*
Password="PASSWORD"	*Only enter a password if it is required for your network*
Debug=0	*Assigned by Config software*
Id=5300xxx	*Your **DMR ID** allocated on the config page*
Location=1	*Assigned by Config software*
Name=DMR+_IPSC2-NZ-HOTSPOT	*The DMR+ Server you chose in the config*

Options="StartRef=4850;RelinkTime=15;UserLink=1;TS1_1=1;TS1_2=13;TS1_3=5;TS1_4=235;TS2_1=505;" *You set this in DMR Options on the Config page. It can be edited here.*

You may have noticed that there are three quite complicated TGRewrite lines. I use this structure because I transmit to the hotspot on TS2 but the IPSC2 DMR network server has some static talk groups on TS1. It is exactly the situation you will encounter if you have a simplex hotspot with two or more DMR networks connected.

Line one, TGRewrite1=2,8000001,1,1,503 translates talk groups 1 to 504 from TS2 on the radio to TS1 on the DMR+ network. It also offsets the talk groups from 8000001 – 8000504 to 1 – 504.

TGRewrite2=2,8000505,2,505,45 translates 8000505 - 8000550 on TS2 to 505 - 550 TS2. This time there is no conversion to TS1 because, on the IPSC2 server, TG 505 and TG 530 are static on TS2. The third line sends the remaining talk groups to TS1.

Click **Apply Changes** to save the changed TG values.

ADDING THE TGIF NETWORK

You must register to use the TGIF network. It is a simple process involving filling in an online form and verifying your email address. After the application has been accepted you will get a second email welcoming you to the network.

Register for TGIF network at https://prime.tgif.network

There is a handy video on this topic by Andy Neilson G7LRR at https://www.youtube.com/watch?v=I9DgTEbGx_E

WARNING: To add a second or third DMR network you need to get into the 'Expert' area of Pi-Star and edit the config files. I am not a Pi-Star expert, so I will explain the settings that work for me. There may be other methods that work just as well, or better. I have not attempted to learn the hundreds of other settings, so play with them at your own risk.

You MUST have a working configuration before you attempt to add a second or third DMR network! Things can only get worse if you do not.

The very first thing you should do is **make a backup** so that you can get back to a working configuration if everything goes "pear-shaped." See the Pi-Star backup section on page 184.

Talk group offsets

Just like when we added the DMR+ network, we will need to make some 'Expert' changes to shift the TGIF talk group range so that the hotspot knows which network to send the calls to.

This time we are going to use an offset of 5000000. That means to use TGIF talk group 110 we make a talk group 5000110, create a channel on our hotspot frequency in the usual way and add the channel to a zone, just like normal.

The hotspot subtracts 5000000 from the TG number and sends the call to the TGIF network as TG 110. A call originating on talk group 110 on the TGIF network gets 5000000 added before the call is transmitted by the hotspot.

Adding the TGIF network

There is no space on the Configuration page for adding the TGIF network. Network 1 is reserved for Brandmeister, and Network 2 is reserved for DMR+. You do not have to enable network two if you do not want the DMR+ network, but you cannot use the Network 2 config area for another network. You have to use Network 3 or 4.

Click **Configuration > Expert > Full Edit: DMR GW**. This opens the screen that writes one of the Pi-Star initialisation files.

Scroll down the list until you see [DMR Network 3]. There is a list of initialisation statements that have resulted from the settings we changed so far. Check out the meanings of these statements in the 'Adding a DMR+ network' section above.

A real example – network 3 (TGIF)

[DMR Network 3]

Enabled=1

Name=TGIF_Network

PCRewrite1=1,5009990,1,9990,1

PCRewrite2=2,5009990,2,9990,1

TypeRewrite1=1,5009990,1,9990

TypeRewrite2=2,5009990,2,9990

TGRewrite1=1,5000001,1,1,999999

TGRewrite2=2,5000001,2,1,999999

Address=prime.tgif.network

Password="passw0rd"

Port=62031

Location=0

Debug=0

Id=Your DMR ID number here

Change the DMR Network 3 settings so that they are the same as the example.

Click **Apply Changes**. It is not a bad idea to also reboot the hotspot. Sometimes the enable switches drop off and one or more of the networks fall off. If that happens reset the enable switches to 1 again and **Apply Changes**. They usually stick the second time.

New Password rule

Some TGIF talk groups will still work with the default '**passw0rd**' however since the 16th July 2021 some of the popular talk groups now need a system-generated password. This is obtained by registering with TGIF. After that is completed, log on to the TGIF website and select your **callsign** (top right). On the pop-up list, select **User Security**. You will see your DMR ID which the site finds as a part of the registration process and a Hotspot Security Key.' Copy the security key to your PC clipboard by clicking the blue clipboard icon, or CTRL-C. Go to your TGIF configuration on your Pi-Star dashboard and paste the security key in between the double quotes on the Password line. Password="**ABCDEFG5123456789**".

Static talk groups

Static talk groups work a little differently on the TGIF network. Keying up a TGIF talk group on either time slot makes it auto-static.

The time slot will stay allocated to your hotspot until you key up a channel with a different talk group. You can also set the talk groups that will be sent to your hotspot in the Selfcare page on the TGIF dashboard. Click **your callsign** (top right) and select **Selfcare**. Enter a talk group name or number or select one using the dropdown list. With a simplex hotspot, you can only select a talk group for TS2.

Figure 67: TGIF static talk groups

TIP use 'last heard' on the TGIF dashboard to see what talk groups are active.

TIP: the TGIF Parrot does not work via the Brandmeister gateway. It works fine if your hotspot is only connected to TGIF but not if TGIF is a second or third DMR network connected via the Gateway. The channel will key up and you can see it on the TGIF 'last heard' screen, so it is still useful as a check that you are accessing the network, but you do not get your audio repeated back.

ADDING A DIFFERENT NETWORK

If you want to add a different network such as FreeDMR, FD, or the HB network as your second or third DMR network, you can modify the TGIF settings. They must be placed into Network 3 or Network 4 because Network 1 is reserved for Brandmeister, and Network 2 is reserved for DMR+.

You should only have to change the **Name, Address, Password**, and possibly **Port**. You could change the offset, but if you are not using TGIF then 5000000 will be fine.

If you don't know what to enter, temporarily change DMR master to your network server of choice, as if the hotspot was operating with a single DMR network. Check out the **Name, Address, Password**, and **Port** that the Config page generates **for** [DMR Network 1]. Then return the DMR Master to **DMR Gateway** and use the information for your Network 3 or Network 4 settings.

If you want to use FreeDMR, FD, or the HB network as the only network connected to the hotspot, follow the instructions back on page 180.

MCC

Have you wondered how the DMR ID and talk group numbers are allocated? They conform to an international telecommunications agreement, ITU-T Recommendation E.212, which defines 'Mobile Country Codes,' (MCC).

2 = Europe

3 = North America and the Caribbean

4 = Asia and the Middle East

5 = Oceania (Pacific, Australia, New Zealand)

6 = Africa

7 = South and Central America

9 = Worldwide services such as satellite, aircraft, maritime, and Antarctica

Countries are allocated three-digit prefixes based on the regional codes, and the DMR system uses those numbers as the basis for most talk group and DMR ID numbers.

For example, the UK is 234-235, the USA is 310-316, Canada is 302, New Zealand is 530 and Australia is 505. Talk group 3026 is a Canadian talk group for Alberta, and Private DMR ID 5302030 is the Wellington VHF Group in New Zealand.

Mobile country code	Country	Alpha
213	Andorra	AD
722	Argentina	AR
283	Armenia	AM
505	Australia	AU
232	Austria	AT
400	Azerbaijan	AZ
426	Bahrain	BH
470	Bangladesh	BD
206	Belgium	BE
362	Bonaire, (Curacao)	BQ
218	Bosnia and Herzegovina	BA
724	Brazil	BR
284	Bulgaria	BG
302	Canada	CA
730	Chile	CL
460	China	CN
732	Colombia	CO
712	Costa Rica	CR

219	Croatia	HR
368	Cuba	CU
362	Curaçao	CW
280	Cyprus	CY
230	Czech Republic	CZ
238	Denmark (Kingdom of Denmark)	DK
370	Dominican Republic	DO
740	Ecuador	EC
602	Egypt	EG
706	El Salvador	SV
248	Estonia	EE
244	Finland	FI
208	France	FR
647	French Indian Ocean Territories (France)	RE
202	Greece	GR
352	Grenada	GD
704	Guatemala	GT
372	Haiti	HT
708	Honduras	HN
454	Hong Kong	HK
216	Hungary	HU
404	India	IN
510	Indonesia	ID
272	Ireland	IE
234	Isle of Man (United Kingdom)	IM
425	Israel	IL
222	Italy	IT
338	Jamaica	JM
440 - 441	Japan	JP
401	Kazakhstan	KZ
450	Korea, South	KR
247	Latvia	LV
415	Lebanon	LB
295	Liechtenstein	LI
246	Lithuania	LT
270	Luxembourg	LU
294	North Macedonia	MK
502	Malaysia	MY
278	Malta	MT
334	Mexico	MX
259	Moldova	MD

297	Montenegro	ME
604	Morocco	MA
204	Netherlands	NL
530	New Zealand	NZ
710	Nicaragua	NI
242	Norway	NO
714	Panama	PA
716	Peru	PE
515	Philippines	PH
260	Poland	PL
268	Portugal	PT
330	Puerto Rico (United States of America)	PR
427	Qatar	QA
226	Romania	RO
358	Saint Lucia	LC
292	San Marino	SM
420	Saudi Arabia	SA
220	Serbia	RS
525	Singapore	SG
362	Sint Maarten	SX
231	Slovakia	SK
293	Slovenia	SI
655	South Africa	ZA
214	Spain	ES
240	Sweden	SE
228	Switzerland	CH
520	Thailand	TH
374	Trinidad and Tobago	TT
286	Turkey	TR
255	Ukraine	UA
430	United Arab Emirates (Abu Dhabi)	AE
234 - 235	United Kingdom	GB
310 - 316	United States of America	US
748	Uruguay	UY
541	Vanuatu	VU
734	Venezuela	VE
452	Vietnam	VN

Troubleshooting

RADIO KEYPAD BUTTONS NOT WORKING

The keypad may be locked. See your radio manual for details on how to unlock it. On my radio, I had to press the Menu (-) key and the Star (*-) key at the same time to unlock the radio.

CANNOT PROGRAM THE RADIO FROM THE RADIO KEYPAD

Many radios are designed for commercial radio as well as amateur radio use. You cannot use the radio keypad buttons to change the channel frequencies, talk groups, or other configuration settings if the radio is set to the "Professional Mode." This is because commercial DMR network operators do not want the (unlicensed) radio users changing the configuration. You need to use the radio configuration software to change the 'Working Mode' setting from 'Professional Mode' to 'Amateur Mode.' On my Radioddity CPS it is under, **Common Setting > Optional Setting > Work Mode > choose working mode > amateur mode.**

On the TYT MD-UV390 CPC try **Menu Item** and check the **Program Radio** checkbox in the **Utilities** area.

OUTPUT POWER IS LOW

Some manufacturers are prone to exaggerating the RF output power. More specifically the radio probably will transmit at the stated power in FM mode if the battery is at full charge, but the output power may be much lower if the battery is down a bit. **Don't worry about it!** Remember that transmitting at half power is only a 3 dB reduction in signal strength at the repeater. So, unless you are right on the fringe, running slightly less transmitted power will not make any difference. Use the lowest power setting if you are transmitting to a hotspot. You don't need full power to get across the room. For repeaters, use the lowest power setting that works reliably. Using the radio at full power makes it run hot and dramatically increases battery consumption.

The average power will read lower on DMR because the radio transmits on one time slot and then turns off during the other time slot.

DISPLAY THE CHANNEL NAME INSTEAD OF THE FREQUENCY

It is usually preferable to display the channel name instead of the frequency. On my setup it is under, **Common Setting > Optional Setting > Work Mode > Display mode > Channel**, (or optionally Frequency).

TIP: Of course, you have to save the edited code plug back to the radio before the changes take effect. Always save a backup copy of the new configuration on your PC as well.

SEARCHING PDF MANUALS

The PDF search function does not work on the downloaded manual for the Radioddity GD-AT10G. But the manual for the AnyTone AT-D878UV, which is an almost identical radio, is fully searchable.

CAN'T HEAR THE REPEATER

You have the DMR channel programmed, and you can see transmissions on the RSSI (receive signal strength indicator), but you cannot hear anything. There are several possible explanations. Firstly, if the repeater is a 'multi-mode repeater.' It could be transmitting an FM, P25, D-Star, or Fusion mode signal. You will only be able to hear the transmission if it is a DMR signal.

If it is transmitting a DMR signal, you may be listening to a different talk group, or you might have the wrong time slot or colour code selected. Use the 'digital monitor' (sometimes called promiscuous mode) to find out the correct talk group, colour code and time slot for the repeater and change the channel setting to match.

On the Radioddity/AnyTone CPS, you can turn on the 'digital monitor in the CPS. Select **Common Setting** > **Optional Setting** > **Digital Func** > **Digital Monitor** and select **Double Slot**. On the same page, set **Digital Monitor CC** to 'Any' and **Digital Monitor ID** to 'Any.' When a call is made the display will show the TG that is in use, the TS (time slot) and the CC (colour code). Once those settings have been made you can click the **PF2** button once to turn on single time slot monitoring and a second time to turn on dual time slot monitoring, the third click turns the monitor off.

On the TYT the button below the PTT should turn on the monitor. If not you can use the CPS, to select **Buttons Definitions** > **Radio Buttons** and use the dropdown list to allocate **Monitor** to either of the **Side Buttons**. Set **General Settings** > **Monitor Type** to **Squelch**.

CAN'T HEAR UNKNOWN ID

If you can hear a known ID (name or callsign comes up on the display) but you cannot hear an unknown ID, the radio is probably set to only accept known IDs. This is a feature designed to stop commercial users from hearing traffic destined for another call group.

On the TYT MD-UV380/390, press **MENU**, scroll down to **Utilities** > **Radio Settings** > **Group Call Match** and select **Turn Off** and **CONFIRM**. I guess you might as well do the same for Private ID calls. Press **MENU**, scroll down to **Utilities** > **Radio Settings** > **Private Call Match** and select **Turn Off** and **CONFIRM**.

I am not aware of a similar setting in the Radioddity GD-AT10G or the AnyTone AT-D878UV.

HEARING EVERYTHING ON THE TIME SLOT

If you are hearing everything on the repeater or hotspot and not just the selected channel, there are several possibilities.

Cause of problem	TYT	AnyTone / Radioddity
Monitor turned on	▶ left facing speaker icon	◀) or ◀)) symbol
Receive group	Group list selected	Receive group list selected
Group Call Match	Off	N/A

1. You may have the monitor turned on (digital monitor on the AnyTone and Radioddity transceivers). On the TYT radio, this is usually activated by 'side button 2' below the PTT. It is indicated with a left-facing speaker icon ▶ between the RSSI indicator and the RF power indicator on the display. Turn off the icon and you should not hear the repeater unless you have a different problem. On the Radioddity transceiver, it can be selected in the CPS or activated by the 'PF2' button below the PTT. It is indicated with a red right-facing speaker icon to the right of the RF power indicator on the display. Turn off the icon and you should not hear the repeater unless you have a different problem. A ◀) symbol means the radio is monitoring all traffic on the same time slot as the selected channel. A ◀)) symbol means the radio is monitoring all traffic on both time slots.

2. You may have a receive group selected, on the channel you are listening to. Receive groups allow you to listen to the selected channel and other nominated channels on the same time slot as the selected channel. This can be annoying and confusing.

   TIP: For repeaters and duplex hotspots, I recommend that you create a copy of your favourite channel that is using a talk group on TS1. Name it TS1 Monitor. Associate a receive group containing all the TS1 talk groups that you would like to monitor.

   For all other channels do not select a 'Receive group list' or 'Group list.'

   That way if you select a channel, you will only hear traffic on that channel. If you see a signal come up on the RSSI and LED, you can pop the monitor on to hear and see what it is. If you want to monitor, select the 'TS1 or TS2 monitor' channel.

3. The TYT radio has a menu option called 'Group Call Match.' It should be turned on, otherwise, the radio ignores the group call IDs and you hear everything. **Menu > Utilities > Radio Settings > GroupCallMatch > Turn On.**

TYT RADIO CANNOT SELECT THE B (LOWER) CHANNEL

Symptom: You can't select the lower channel and the Menu only has 'Zone' on it. This is caused by the radio starting up in a zone that has no B channels. When the radio boots after a code plug update it will start on the zone that is at the top of the Zone list. If that zone has no B channels, the radio will not let you select the B channel. If you can't select the B channel, you cannot change to a zone that has B channels. I expect the same problem would occur if the first zone had no A channels. Make sure that every zone has at least one A channel and one B channel.

REPEATER NOT FOUND

A Repeater not found message indicates that the repeater you have selected did not respond. It usually means that it is out of range or already transmitting a different talk group. Or it could be handling a call from a YSF or D-Star radio. Try again when the RSSI reading on your radio indicates that the repeater is not transmitting.

???? ON SETUP SCREEN

My setup screen had ???? on three of the options and a spelling mistake on 'Disable all LEDS.' I have relabelled them below. Group Call Match should be selected.

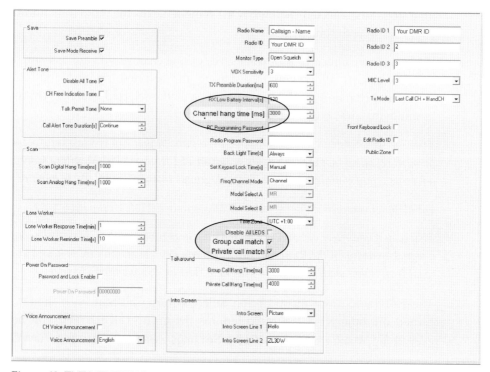

Figure 68: TYT MD-UV380 setup screen

TURN ON GPS AND APRS

On the AnyTone and Radioddity CPS, select **Tool** > **Options** and check the **GPS** box. On the AnyTone and Radioddity CPS, select **Tool** > **Options** and check **APRS**.

SCAN FUNCTION NOT WORKING

Symptoms: The **Menu** > **Scan** selection on the radio does nothing and pressing the relevant side panel function key to start the radio scanning doesn't work either, yet you know you have at least one scan group, and you have channels allocated to the scan group. The answer is not documented anywhere, but it is quite simple. You cannot initiate a scan or enter the Scan Menu unless the radio is set to one of the channels that have a scan group associated with it.

HIGH BER INDICATED ON PI-STAR

High BER when you transmit shows up as a red indicator on the Local RF Activity and Gateway Activity screens on Pi-Star, and also on the Brandmeister, TGIF, or IPSC2 server dashboard. It usually means that your hotspot is off frequency and an offset needs to be applied in Pi-Star (not in your radio!). See 'Hotspot off frequency,' on page 205 below.

MY NEW HOTSPOT DOESN'T WORK

Don't panic! You are much more likely to have a problem with a simplex MMDVM hotspot than a duplex MMDVM hotspot. Other hotspots like OpenSpot, DV dongles, etc. are less prone to problems and are better supported by the manufacturer. The most common problems are,

- An insufficient power supply. Symptoms, include Windows 'bonging' regularly, hotspot rebooting, no display or incorrect display.

- Display not configured correctly in Pi-Star. No display even when rebooting the hotspot.

- Radio on the wrong zone (I was most confused when trying to transmit on my other hotspot's frequency and zone).

- Incorrect colour code, frequencies, time slot, or zone configuration.

- Incorrect Modem setting in Pi-Star. No transmission to the radio. No display.

- Network disabled in Pi-Star. Check the bottom right box on the Pi-Star dashboard. The Pi-Star software has a bad habit of turning networks off, particularly network 3 and network 4.

- Hotspot being well off frequency. Seeing activity on the hotspot display but no audio from the radio, no display on the radio of calling stations, hotspot not responding when you transmit.

Insufficient power supply

An insufficient power supply can be a problem, especially if you are using a Raspberry Pi 3 rather than a Raspberry Pi Zero W. I have had no issues powering my hotspot from a USB3 port on my computer, but the recommendation from the Raspberry Pi Foundation is to use a 2.5 amp USB power supply or a USB battery bank.

Display configuration

The display might not work if there is insufficient power, or it is not configured correctly in Pi-Star.

For an OLED (small display) select OLED type 3 for the very small 0.96' screen or OLED type 6 for the more common 1.3" screen. Either works on my hotspot. Port should be set to select /dev/tty/AMA0 because the display is being driven directly by the modem board.

If you have a Nextion screen, select Nextion. The port is usually set to modem unless the display is connected to a TTL to USB adapter plugged into the Rpi. In that case select /dev/tty/USB0.

Wrong zone

Check that you are in the right zone on the right channel with the correct frequencies. It is easy to get mixed up, particularly if you have two hotspots.

Incorrect colour code, frequencies, time slot, or zone

Check that your channels match the DMR colour code setting in the DMR Configuration section of Pi-Star. All channels on a simplex hotspot are on TS2. Use the digital monitor to listen for any traffic on the time slot. Check your talk group matches the channel and is in a zone. If the network needs the call to be on TS1 and you are using a simplex hotspot, check the TGRewrite commands in Pi-Star.

Incorrect Modem setting in Pi-Star.

This is indicated by no transmission to the radio and/or no hotspot display. Check the advertisement for the modem you have purchased and see if there is any indication of the modem setting that should be set in Configuration > General Configuration > Radio/Modem Type. My duplex modem is set to MMDVM_HS_Hat_Dual Hat (VR2VYE) for Pi (GPIO), as specified by the manufacturer.

Dual Hat means it is a duplex modem, VR2VYE is the person who wrote the firmware, it is for a Pi, and it is connected to the Pi using the GPIO header pins, not a USB interface.

I had trouble selecting the correct modem for my simplex hotspot, mostly because it was so far off frequency it was not working anyway.

There is a selection of choices. Make sure that you choose a Pi version, rather than a DV-Mega, Nano or NPi version. Unless you do have one of those.

The **STM32-DVM / MMDVM_HS – Raspberry Pi Hat (GPIO)** modem seems to work with most generic Chinese simplex hotspots. Some of the other choices displayed calls on the hotspot, but there was no signal received by the radio.

Network disabled in Pi-Star

Check the bottom right box on the Pi-Star dashboard. The software has a bad habit of turning networks off, particularly network 3 and network 4. Check the two 'network enable' switches on the configuration page. Turn on the network or networks you want to use. Check the DMR mode switch is also turned on. Then click '**Apply Changes.**' If you are using network 3 or 4, go into **Expert > Full Edit DMR GW** and scroll down to network 3 or 4. Set Enabled = **1**. Click **Apply Changes** at the bottom of the Expert page. Do not click Apply Changes on the Configuration page.

Hotspot off frequency

This is the one that trips up most people, probably because it is the hardest to fix. It is very common for a new hotspot to be off frequency. The frequency error should be corrected in Pi-Star and never with the radio or CPS offsets. There is no point in deliberately making your radio transmit off frequency to compensate for the frequency error in the hotspot. How do you know it is the hotspot, not the radio? It is far more likely to be the hotspot. The only way to check for sure would be to compare it with another radio.

TIP: Some hotspots are checked by the vendor and shipped with an offset number on a slip of paper in the box or stuck to the bottom of the modem. I was not lucky in that respect.

When I powered up my new simplex hotspot it would not respond to my handheld radio when I transmitted and although I could see callers on the hotspot display and a LED and RSSI indication on the radio, I could not hear them. Even with the digital monitor turned on. OK, I did panic!

It wasn't until I watched a setup video on YouTube, I remembered the hotspot frequency offsets. I am fortunate to own a spectrum analyser and a good frequency counter, so I was able to quickly find out that my hotspot was a whopping 4.4 kHz off frequency. The online video was talking about a frequency offset of only 472 Hz.

If you don't have a frequency counter or spectrum analyser, you can use an SDR receiver. I do not recommend a direct connection. Attach an antenna to the SDR. The frequency accuracy of your SDR should be good enough to determine if the modem is transmitting a long way off frequency. If you don't have an SDR either, you can use the 'trial and error' method. Set a 500 Hz offset and see if you can hear traffic on the hotspot or if it displays your callsign when you transmit to it.

Keep stepping up (or down) in 500-hertz steps until you hopefully have some success. I think it is unlikely the hotspot will be more than 5000 Hz high or 5000 Hz low.

Setting the frequency offset

You can be pretty sure that any offset applied to the hotspot's transmit frequency will also be correct for its receive frequency. So, always set both offsets the same.

There are two ways you can do this. Either adjust the receiver offset for the best BER (bit error rate) when you transmit to the hotspot. Or adjust the transmit offset while the hotspot is transmitting, by measuring the frequency it is radiating with a spectrum analyser, frequency counter or SDR receiver. If the hotspot is seeing the transmission from your radio and the dashboard is showing 'Local RF Activity' use option 1. If the hotspot is not seeing the transmission from your radio and the dashboard is not showing 'Local RF Activity,' but is showing 'Gateway Activity,' use option 2. Option 2 is the most accurate, but it requires test equipment. Option 1 is a perfectly acceptable method.

Option 1: adjust the receiver offset for best BER

This method only works if the hotspot is seeing your transmission and it is showing up on the 'Local RF Activity.' If it is not and you do not have a frequency counter, spectrum analyser or SDR receiver capable of receiving the 70cm band you can try transmitting at a range of offsets until you can get into the hotspot. Make very sure that the hotspot's receiver frequency is the same as your radio's transmitter frequency.

Open Pi-Star and look at the 'Dashboard' page. Specifically, the BER (bit error rate) indication in the 'Local RF Activity' section.

1. Set your radio to a channel (on the hotspot) that won't annoy people. The Parrot (Echo) talk group 9990, or BM TG 98 (Test), or TG 9 (Local), or TG 4000 Disconnect are all good options.

2. Key up your radio with the PTT and hold the transmit for four or five seconds. You should see your callsign pop up and an indication on the BER meter. If it is green the frequency offset is good. Job done. If it is red the frequency offset needs to be adjusted.

   To adjust the offset in Pi-Star, select Configuration > Expert > MMDVM Host. Scroll down to the Modem section and find RXoffset and TXoffset. (Not RXDCoffset or TXDCoffset).

3. Set both RXoffset and TXoffset to plus 500 Hz. Click Apply Changes.

4. Transmit again after the hotspot reboots and note if the BER got better or worse.

a. If it got worse try a -500 Hz offset and try again.

b. If it got better, increase the offset and try again. Keep changing both offsets until the BER is less than 0.5%.

-//-

Option 2: adjust the transmitter offset

Using a frequency counter or spectrum analyser is more accurate than using the BER method, but it does require you to have the test equipment.

Remember to use a 20 dB attenuator in the cable if you are connecting your hotspot to a frequency counter or spectrum analyser.

SDR receivers are very sensitive. They are designed to receive signals down to around -100 dBm. **I do not recommend a direct connection between the hotspot and an SDR receiver.** Connect the SDR to an antenna. If you do decide to use a direct connection, you will need a 40 dB or 50 dB attenuator in the cable between the radio and the SDR.

1. If possible, make a busy channel static so that there is plenty of activity on the channel.

2. Take a note of the hotspot's transmitter frequency (or simplex frequency). It is in the Radio Info box on the Pi-Star dashboard.

3. In the Pi-Star software select **Configuration > Expert > MMDVM Host**. Scroll down to the Modem section and find **RXoffset** and **TXoffset** (not RXDCoffset or TXDCoffset).

4. Monitor the hotspot and observe the frequency when it transmits. Subtract the wanted frequency as indicated on the Pi-Star dashboard from the observed frequency.

5. Adjust both the **RXoffset** and **TXoffset** by the amount of offset needed in Hertz. Click **Apply Changes**. After the hotspot reboots, observe the frequency when the hotspot transmits. Make any minor adjustments. You should be able to get to within about 100 Hz. The adjustment process is not fine enough to get the frequency exactly right.

6. You should make a final check that the BER is good when you transmit to the hotspot, using the instructions in option 1: The BER indication should be less than 0.5% and green.

DMR NETWORK 3 KEEPS TURNING OFF

On my system using the **Apply Changes** button on the Configuration page always turns off network 3 (TGIF). I had to go and enable it again in the Expert section. Then click the **Apply Changes** button on the Expert page, not the Configuration page.

Glossary

59	Standard (default) signal report for amateur radio voice conversations. A report of '59' means excellent readability and strength.
73	Morse code abbreviation 'best wishes, see you later.' It is used when you have finished transmitting at the end of the conversation.
.dll	Dynamic Link Library. A reusable software block that can be called from other programs.
2m, 70cm	Two metre (144 MHz) and 70cm (430 MHz) amateur radio bands
4FSK	DMR transmitters use 4 state frequency shift keying modulation. Each frequency shift carries two bits of the input data stream.
A/D	Analog to digital
ADC	Analog to digital converter or analog to digital conversion
AF	Audio frequency - nominally 20 to 20,000 Hz.
Algorithm	A process, or set of rules, to be followed in calculations or other problem-solving operations, especially by a computer. In DSP it is a mathematical formula, code block, or process that acts on the data signal stream to perform a particular function, for example, a noise filter.
AMBE	Advanced Multi-Band Excitation. Vocoder used for digital voice on Inmarsat and Iridium satellite phones, D-Star, and 'phase 1' versions of P25 digital.
AMBE+2	The AMBE+2 Vocoder uses a propriety chip made by DVSI (Digital Voice Systems Incorporated) to convert speech into a coded digital signal or the received digital signal back to speech. Can transmit intelligible speech with data rates as low as 2 kbs. Used for DMR, YSF, NXDN, and 'phase 2' P25.
APCO	Association of Public-Safety Communications Officials
APRS	Automatic packet reporting system – used to send and display location information from a GPS receiver. APRS beacons transmitted over DMR are displayed on the APRS.fi website.
BER	Bit error rate – a quality measurement for any digital transmission system. It measures the number of bits that were received incorrectly compared to the overall bit rate.
Bit	Binary value 0 or 1.
Brandmeister	Brandmeister operates the largest DMR network. Currently 1606 Talk groups. The Brandmeister dashboard is showing 4982 connected repeaters and 14,154 connected Hotspots.

BW	Bandwidth. The range between two frequencies. For example, an audio passband from 200 Hz to 2800 Hz has a 2.6 kHz bandwidth.
C4FM	Continuous 4 state Frequency Modulation (used for P25 Phase 1 and Fusion)
Carrier	Usually refers to the transmission of an unmodulated RF signal. It is called a carrier because the modulation process modifies the unmodulated RF signal to carry the modulation information. A carrier signal can be amplitude, frequency, and/or phase-modulated. Then it is referred to as a 'modulated carrier.' An oscillator signal is not a carrier unless it is transmitted.
c-Bridge	Developed by Rayfield Communications for Motorola. The c-Bridge links local networks together into nationwide and ultimately worldwide networks.
code plug	A code plug is the radio's configuration file. It is the data file that the programming software loads into the radio. Sometimes the data can also be changed using the front panel buttons.
CODEC	Coder/decoder - a device or software used for encoding and decoding a digital data stream.
Colour Code	The colour code is used to identify the output of a specific repeater, much like the CTCSS tone on an FM repeater. In the unlikely event that two repeaters near you are using the same frequencies. Using different colour codes will ensure that you only hear the wanted repeater.
CPS	Customer Programming Software is another name for radio configuration software.
CPU	Central processing unit. The ARM (advanced RISC machine) processor in the Raspberry Pi, or the microprocessor in your PC. [RISC is reduced instruction set computing, an acronym inside an acronym.]
Cross-connect	A link between different technologies or networks. Such as a Brandmeister Talk Group linked to a DMR-MARC Talk Group. Or a Brandmeister Talk Group linked to a System Fusion 'Room' or 'Reflector'
CTCSS	Continuous Tone Coded Squelch System, used for access control to most analog FM repeaters and FM handheld or mobile radios.
D/A	Digital to analog.
DAC	Digital to analog converter or digital to analog conversion
Dashboard (DMR network)	A network dashboard is an HTML website that displays the status of a DMR network server or a complete DMR network. The Brandmeister and TGIF dashboards provide information on the whole network, including live 'last heard' updates. In the case of the

	DMR+ network and regional DMR networks, there is an IPSC2 dashboard for each server.
Dashboard (repeater)	A repeater dashboard is an HTML website that displays the status of a repeater or Hotspot, including what frequencies it is on, its location, who owns it, who is using it, and what talk groups it is linked to. Most repeater dashboards are available to the amateur radio community, or selected people, over the Internet. You can choose whether to make your Hotspot public or private.
data	A stream of binary digital bits carrying information
dB, dBm, dBc, dBV	The Decibel (dB) is a way of representing numbers using a logarithmic scale. Decibels are used to describe a ratio, the difference between two levels or numbers. They are often referenced to a fixed value such as a Volt (dBV), a milliwatt (dBm), or the carrier level (dBc). Decibels are also used to represent logarithmic units of gain or loss. An amplifier might have 3 dB of gain. An attenuator might have a loss of 10 dB.
DC	Direct Current. The battery or power supply for your radio, charger, or Hotspot will be a DC power supply.
Direct Mode	is another name for a direct radio to radio contact on a simplex frequency.
DMR	For many years DMR stood for Digital Microwave Radio, used for linking Cell Sites and radio stations, telephony nodes, TV transmitters, data centres, and many other technologies. I don't know how the mobile radio guys managed to steal the acronym.
DMR	If you don't know this, you haven't been paying attention. DMR stands for Digital Mobile Radio. A standard for sending voice traffic over a digital radio link to another radio, through a repeater, or to a station connected to a connected Talk Group.
DMR+	DMR+ is a worldwide DMR network. It was the first to interconnect ETSI standard Tier II repeaters. It is aligned with the DMR-MARC network so you can access the DMR-MARC Talk Groups. The DMR+ network specialises in interconnections with other technologies such as D-Star, AllStar, and C4M (P25 and YSF).
DMR-MARC	DMR-MARC is a network of DMR repeaters established by the Motorola Amateur Radio Club. The members of MARC were instrumental in getting DMR established for amateur radio. They set up the first amateur radio DMR networks and repeaters. There are around 500 DMR-MARC repeaters in 83 countries with over 144,000 registered users.
D-Star	Digital Smart Technologies for Amateur Radio. D-Star is the (mostly) Icom digital voice system. Unlike DMR it was developed specifically for amateur radio.

DVSI	The AMBE+2 Vocoder uses a propriety chip designed and made by DVSI (Digital Voice Systems Incorporated) to convert speech into a coded digital signal or the received digital signal back to speech.
Duplex	A radio or Hotspot that can receive and transmit at the same time. Usually on different frequencies. A standard repeater is a duplex system.
DX	Long-distance, or rare, or wanted by you, amateur radio station. The abbreviation comes from the Morse telegraphy code for 'distant exchange.'
Echo	A Talk group that repeats back a test transmission that you make. The same as Parrot.
ESSID	Extended service set identifier – a 2 digit extension to your DMR ID number to identify a second or subsequent hotspot on the same network.
ETSI	The European Telecommunications Standards Institute developed the DMR platform.
FM	Frequency modulation. The "good ol'" analog repeater system.
FSK	Frequency Shift Keying. DMR transmitters use 4 state frequency shift keying modulation. Each frequency shift carries two bits of the input data stream
FTDI	USB to 3.3V TTL level converter designed by Future Technology Devices International Ltd.
GMSK	Gaussian Minimum Shift Keying - spectrum efficient frequency shift keying mode (D-Star)
GPS	Global Positioning System. A network of satellites used for navigation, geolocation, and very accurate time signals.
Hex	Hexadecimal – a base 16 number system used as a convenient way to represent binary numbers. For example, 1001 1000 in binary is equal to 98h or 152 in decimal.
Hotspot (DMR)	A DMR Hotspot is a small Internet connected box that can connect to DMR Talk Groups. You transmit from your DMR handheld to the Hotspot, and it passes the data through to the Internet. The information that is returned is transmitted by the Hotspot back to your radio. 77% of all DMR Hotspots are MMDVM. The rest are mostly OpenSopt, DVMega, Motorola, and Hytera Hotspots.
Hotspot (WiFI)	Many cell phones can be configured to act as a WiFi hotspot, allowing WiFi devices to get access to the Internet via your phone and mobile data plan. You could connect a WiFi enabled DMR Hotspot (OpenSpot3) to a WiFi Hotspot on your phone and connect your DMR to worldwide Talk Groups via your phone.
Hz	Hertz is a unit of frequency. 1 Hz = 1 cycle per second.

IMBE	Improved Multi-Band Excitation. Vocoder used for P25 phase 1 digital voice
IPSC	Internet Protocol Site Connect is a series of different protocols used to connect repeaters within a specific network.
kHz	Kilohertz is a unit of frequency. 1 kHz = 1 thousand cycles per second.
LAN	Local Area Network. The Ethernet and WIFI connected devices connected to an ADSL or fibre router at your house are a LAN.
LED	Light Emitting Diode
Linked in DMO mode	Simplex Hotspots are linked in DMO (Direct Mode Operation) mode. It means the mode which two DMR radios use to talk directly radio to radio. In other words, it is a fancy way of saying Simplex operation with a network connection.
LoTW	Logbook of the World. An ARRL QSO logging database used worldwide.
MCC	Mobile country code – used for allocating DMR ID and talk group numbers.
MHz	Megahertz – unit of frequency = 1 million cycles per second.
MIC	Microphone
MMDVM	Multi-mode digital voice modem - usually supports DMR, D-Star, Fusion, P25, and NXDN. 77% of all DMR Hotspots are MMDVM Hotspots. The rest are mostly OpenSopt, DVMega, Motorola, and Hytera.
MOTOTRBO	MOTOTRBO is a Motorola trademark used to describe their range of DMR products
Network (DMR)	A DMR network is a collection of interconnected repeaters and Hotspots. There are many independent amateur radio DMR networks. The most popular are Brandmeister, DMR+, and TGIF.
Onboard	A feature or data list that is contained within the radio.
Parrot	A colourful class of birds known for their ability to mimic speech and other sounds. In DMR it is a system that repeats back a test call made from your DMR. Also known as Echo.
PC	Personal Computer. For the examples throughout this book, it means a computer running Windows 10.
PSK	Pre-shared key – a password or security code known to your router and a connected router. For example, your BM password.
PTT	Press to talk - the transmit button on a microphone – pressing the PTT makes the radio transmit.
QSO	Q code – an amateur radio conversation or "contact."
QSY	Q code – a request or decision to change to another frequency.
RF	Radio Frequency

RS232	A computer interface used for serial data communications.
RPi	Raspberry Pi single board computer
RX	Abbreviation for receive or receiver
Simplex	Simplex means to receive and transmit on the same frequency. In most cases, a radio operating on a simplex frequency cannot transmit and receive simultaneously. Simplex can be used if you wish to communicate directly with another DMR radio (without a repeater or the Internet).
Simplex Repeater or Hotspot	A simplex repeater uses the same frequency for receiving and transmitting. It passes data (digital voice) from an Internet connection to the hotspot or repeater transmitter so that you can receive it on your radio. When you transmit, the simplex repeater or hotspot passes the signal received from your radio to a talk group over the Internet connection.
Slot 1 and 2 linked	Slot 1 and 2 linked, means that both time slots are linked to the DMR network. Some repeater operators only link one time slot and leave the other for local calls or possibly a connection to a different network. Applies to DMR repeaters and duplex hotspots.
Split	The practice of transmitting on a different frequency to the one that you are receiving on. Repeaters use a 'repeater split' between the repeater input frequency and the repeater output frequency.
Squelch	Squelch mutes the audio to the speaker when you are in FM mode and not receiving a wanted signal. When a signal (with the correct CTCSS tone, if enabled) is received, the squelch 'opens' and you can hear the station.
SSID	Service set identifier – in Pi-Star it is your DMR ID number or your WiFi network name
Tail	Furry attachment at the back of a dog or cat. Also, the length of time an FM repeater stays transmitting after the input signal has been lost. It can also mean a short flexible length of coaxial cable at the antenna or shack end of your main feeder cable.
Talk Group	A Talk Group is the equivalent of a D-Star 'Reflector' or a Fusion 'Room.' It is a collection of linked repeaters that are configured so that users with a common interest or from a common location can talk to each other. For example, there are Worldwide, North American, State, and County Talk Groups. Also, Spanish Language and Old Timers groups.
Talker Alias	Talker Alias is an addition to the DMR specification that adds a user-configurable alphanumeric string to the data stream. It is supported by Brandmeister, SharkRF, and Pi-Star. The obvious use is to send your name and callsign when you transmit, so you don't

	have to have a huge contact database that goes out of date. But many radios don't support the function.
TDMA	Time domain multiple access. A technique for interleaving the data from two or more voice or data channels onto a single data stream. DMR uses TDMA to combine two voice (and data) Time Slots onto a data stream, which is used to modulate the radio using 4FSK.
TGIF	Is the 'Thank God It's Friday' DMR network started in October 2018. It has 726 Talk Groups.
Tier	DMR radios are split into three 'tiers' depending on the capabilities of the radio. Amateur radio DMR uses Tier II.
Time Slot	A DMR repeater (base station) can transmit two voice channels on one RF channel. This means that two people can use the repeater at the same time. The system is called TDMA (Time Division Multiple Access). Each voice channel is carried in a Time Slot. (TS1 or TS2). The repeater transmits a 30 ms burst of data for TS1 followed by a 30 ms burst for TS2. This is repeated until both time slots are idle. A mobile radio transmits a 30 ms burst then turns off for 30 ms, while another user sends their 30 ms burst.
TX	Abbreviation for Transmitting or Transmitter.
UHF	Ultra-High Frequency (300 MHz - 3000 MHz).
USB	Universal serial bus – serial data communications between a computer and other devices. USB 2.0 is fast. USB 3.0 is very fast.
VFO	Variable Frequency Oscillator. Applies to radios that can be tuned in frequency steps rather than stepping through previously saved memory channels.
VHF	Very High Frequency (30 MHz -300 MHz)
Vocoder	A Vocoder is a category of voice codec that analyses and synthesizes the human voice signal for audio data compression, multiplexing, voice encryption or voice transformation. The vocoder was invented in 1938 by Homer Dudley at Bell Labs as a means of synthesizing human speech. [Wikipedia]
W	Watts – unit of power (electrical or RF).
YSF	Yaesu System Fusion

Table of drawings and images

Internet links

The Internet is quite large. I can't mention the thousands of links to all possible DMR references. Here are a few that I found useful.

DMR registration (DMR ID) https://www.radioid.net/

hotspot software and a lot of other reference information https://www.pistar.uk/

Brandmeister https://brandmeister.network/

DMR+ talk groups are at, https://www.pistar.uk/dmr_dmr+_talk groups.php

DMR+ reflectors are at, https://www.pistar.uk/dmr_dmr+_reflectors.php

TGIF talk groups are at, https://www.pistar.uk/dmr_tgif_talk groups.php

The K3NYJ blog has TGIF talk groups at http://k3nyj.blogspot.com/2020/06/talk group-list-tgif-network.html

Brandmeister talk groups are at, https://www.pistar.uk/dmr_bm_talk groups.php

UK talk groups are at, https://www.dmr-uk.net/index.php/layout/

UK Network information https://dmrguide.uk/index.php/dmr-networks/

UK DMR Repeaters & Coverage http://www.ukrepeater.net/dmr.htm

Phoenix IPSC2 dashboard http://phoenix-k.opendmr.net/ipsc/

Phoenix IPSC2 hotspot dashboard http://phoenix-f.opendmr.net/ipsc/

DV Scotland IPSC2 http://dmr1.dvscotland.net/

EditCP software https://www.farnsworth.org/dale/codeplug/editcp/

Contact lists at https://www.radioid.net/ or https://amateurradio.digital/

Free contact list download based on the users of specific BM talk groups at https://brandmeister.network/?page=contactsexport.

TYT MD-UV380/390 firmware update https://www.radioddity.com/pages/tyt-download.

Everything you need to know about DMR in Ireland

https://www.galwayradio.com/wp-content/uploads/2021/09/Digital-Radio-Operating-Manual-v2.pdf

Great videos

Many people have created excellent videos relating to DMR. Sometimes seeing someone perform a task is easier than reading about it. Here are a few that I liked. If you like a video on this list, please hit the 'Like' button and subscribe to the channel. It encourages the creators and helps pay for more great videos.

DMR For Beginners - HAM Radio - TheSmokinApe - YouTube
https://www.youtube.com/watch?v=5FAFt1QCtC0

DMR Explained – Static vs. Dynamic Talk groups – YouTube
https://www.youtube.com/watch?v=uFPA_jFRDKw
DMR Programming Basics - YouTube
https://www.youtube.com/watch?v=Lw0Y-jQZMZ0

DMR for Beginners — How to Connect to Your Local Repeater - YouTube
https://www.youtube.com/watch?v=09YJ0eHRkc8

Complete beginner's guide to DMR Radio! Everything you need to know to get started! – YouTube https://www.youtube.com/watch?v=CE-zXQ1BhSQ

DMR Programming for Amateur Radio - YouTube
https://www.youtube.com/watch?v=a0yzn2rckEo

How to Create a DMR Codeplug 2.0 by Sebastian KBØTTL – YouTube
https://www.youtube.com/watch?v=-COsImyLtGI

[Anytone】 How to Setup Digital APRS on the Anytone 878 - YouTube
https://www.youtube.com/watch?v=kWQqx-z9S-c

How to program a simplex hotspot for multiple DMR networks using Raspberry Pi, Pi-Star & MMDVM – YouTube https://www.youtube.com/watch?v=tMHVzVpCH7U

TYT MD-UV380 / MD-UV380G Review – YouTube
https://www.youtube.com/watch?v=2yds_Jy89FI

DMRGateway BM & TGIF - YouTube
https://www.youtube.com/watch?v=2p6HDJybQQ4

What is DMR Plus? – YouTube https://www.youtube.com/watch?v=ecZbzk6M3SE

Build your own DMR/DStar/Fusion hotspot for CHEAP – YouTube
https://www.youtube.com/watch?v=LspgnvDPJvc

Index

The Author

Well, if you have managed to get this far you deserve a cup of tea and a chocolate biscuit. It is not easy digesting large chunks of technical information. It is probably better to dip into the book as a technical reference. Anyway, I hope you enjoyed it and that it has made learning about DMR a little easier.

I live in Christchurch, New Zealand. I am married to Carol who is very understanding and tolerant of my obsession with amateur radio. She describes my efforts as "Andrew playing around with radios." We have two children and two cats. James has graduated from Canterbury University with a degree in Commerce and is working for a large food wholesaler. Alex is a doctor working in the Christchurch Hospitals.

I am a keen amateur radio operator who enjoys radio contesting, chasing DX, digital modes, and satellite operating. But I am rubbish at sending and receiving Morse code. I write extensively about many aspects of the amateur radio hobby. This is my eleventh Amateur Radio book.

Thanks for reading my book!

73 de Andrew ZL3DW.